Climate Politics as Investment

Simon Wolf

Climate Politics as Investment

From Reducing Emissions to Building Low-carbon Economies

 Springer VS

Simon Wolf
Berlin, Germany

Dissertation an der Universität Kassel, Fachbereich 5: Gesellschaftswissenschaften, unter dem Titel: „Climate politics as investment. A discourse theoretical analysis of how climate change turned into an economic challenge", Disputation am 13. Juni 2012.

ISBN 978-3-658-02405-5   ISBN 978-3-658-02406-2 (eBook)
DOI 10.1007/978-3-658-02406-2

The Deutsche Nationalbibliothek lists this publication in the Deutsche Nationalbibliografie; detailed bibliographic data are available in the Internet at http://dnb.d-nb.de.

Library of Congress Control Number: 2013938485

Springer VS
© Springer Fachmedien Wiesbaden 2013

Printed on acid-free paper

Springer VS is a brand of Springer DE.
Springer DE is part of Springer Science+Business Media.
www.springer-vs.de

Acknowledgements

Writing this book would not have been possible without the support of many people. I am very grateful to my supervisor Christoph Görg, for his advice and critical oversight during all phases of the project. I am also grateful to Peter Newell, for his hospitality in Norwich, and for raising important questions regarding the objectives of my work; to Neil Adger, for welcoming me at the University of East Anglia, and for vital advice and help in orienting my empirical research; to Christoph Scherrer and the participants of his doctoral colloquium in Kassel, for broadening my understanding of discourse theory and the challenges of methodology. The Helmholtz Centre for Environmental Research in Leipzig provided me with a scholarship that allowed me focusing on my research for three years.

In particular, I would like to thank the following people for supporting me in many ways while I was writing this book: Aaron Leopold, Benjamin Stephan, Chris Methmann and Delf Rothe, for contributing to the intellectual environment that helped my project to progress; Martin Bitter and Jakob Horst, for discussing my ideas and questions many times, and for all the talk during coffee breaks; and, of course, Anne Binder and Jenny Jungehülsing, for always being there and cheering me up.

I am most indebted to my family, for constantly supporting me in pursuing my goals, and in particular to you, who in so many ways made me who I am: I would have wanted you to read this book!

Table of contents

Acronyms

AF/AFB	Adaptation Fund/Adaptation Fund Board
AGF	UNFCCC Secretary General High-Level Advisory Group on Climate Finance
AR4	Assessment Report 4 of the IPCC
CDM	Clean Development Mechanism
CIFs	Climate Investment Funds
COP	Conference of the Parties (to the UNFCCC)
CTF	Clean Technology Fund
EU ETS	European Union Emissions Trading Scheme
GDP	Gross Domestic Product
GEF	Global Environment Facility
GGGACC	Global Greenhouse Gas Abatement Cost Curve
GHG	Green House Gases
GIB	Green Investment Bank
GND	Green New Deal
GNDG	Green New Deal Group
IADB	Inter-American Development Bank
IPCC	Intergovernmental Panel on Climate Change
LDCF	Least Developed Countries Fund
MDB	Multilateral Development Bank
MIGA	Multilateral Investment Guarantee Agency
MRV	Monitoring, Reporting, Verification
NAMAs	Nationally Appropriate Mitigation Actions
NGO	Non-Governmental Organisation
PFM	Public Finance Mechanism
REDD	Reducing Emissions from Deforestation and Degradation
SCCF	Special Climate Change Fund
SCF	Strategic Climate Fund
UNEP	United Nations Environment Programme
UNEP FI	UNEP Finance Initiative/Sustainable Energy Finance Initiative
UNFCCC	United Nations Framework Convention on Climate Change
WBCSD	World Business Council on Sustainable Development
WEF	World Economic Forum

1 Introduction

1.1 Towards a new rationality in climate governance

The story of global climate politics in recent years seems to be a story of failure, centred around the widely unsuccesfull efforts to agree on a meaningful follow-up to the Kyoto Protocol. In Kyoto in 1997, the international community had agreed on a framework for global cooperation on climate change, and industrialised countries for the first time in history accepted binding obligations to reduce their carbon emissions. However, the Kyoto Protocol expires at the end of 2012, and despite many attempts in recent years, no breaktrough could be achieved for a similar ambitious treaty.

The peak of this diplomatic tragedy came during the 2009 climate summit in Copenhagen. Climate change had made it to the top of national and international political agendas by then, and the support for an ambitious global climate deal was overwhelming. The biggest number of heads of states ever had announced their presence in the summit, and thousands of climate activitst around the world marched the streets to pressure their governments.

Compared to the enormous expectations in the run-up to the summit, the outcomes of the Copenhagen meeting were widely disappointing. The Copenhagen Accord that was crafted in the last night of the summit by a small group of G20 leaders has no binding legal status and remains far below the ambitions of most national delegations, let alone the expectations of the thousands of NGO campaigners and activitist that gathered in the conference centre and the streets of Copenhagen. Even worse, serious tensions arose between different country groups, who blamed each other for the failure of Copenhagen.

After Copenhagen, the attention for global climate politics fell to a normal level again, and negotiations returned to be the business of experts in national bureacracies. But despite early signs of hope during the first negotiation sessions in 2010 that the tensions created in Copenhagen could be overcome with a new spirit of cooperation, no breakthrough was achieved at the preceeding summits in Cancún 2010 and Durban 2011.

At the time of writing the final version of this book in summer 2012, the future of global climate politics seems to be more uncertain than ever. Even those areas of cooperation on climate change that seemed to have most prospect for progress apparently have lost track during the Durban summit, such as the creation of a REDD mechanism for forest protection and the installation of a Green Climate Fund. Altogether, there is very little evidence that global climate change negotiations will make considerable progress in the next years.

1.1.1 Climate change as economic challenge

Against this background, it seems that the publication of the Stern Review on *The Economics of Climate Change* in 2006 had little lasting effect on climate governance.[1] The study played a central role in climate changes debates in recent years and is widely held as an important factor in pushing the climate issue up the political agenda.

The Stern Review calculates the global macroeconomic costs of climate protection, and compares it with the damage costs of uncontrolled climate change (Stern 2006). It concludes that in the long run, coping with unlimited climate change would be much more expensive than keeping global warming within acceptable limits. Combating climate change, the study argues, makes sense from a purely economic point of view.

Nicholas Stern, the lead author of the study and today one of the most prominent figures in climate change debates, repeatedly calls on governments to raise the ambitions in the climate negotiations. In a book published in 2009, he describes his vision for a Global Deal that would allow to succesfully face the climate challenge while stimulaing a new phase of growth and prosperity.

No doubt, the Stern message was and is strongly taken-up and widely embraced in public and expert debates. It received media attention far beyond what is common for such a study, and even put in the shade the 2007 IPCC Assessment Report in that respect.[2] By many, the study is praised as leading the way towards a new and more succeful approach in combating climate change. Until today, governments, climate experts and activists alike refer to its insights to make the case for an ambitious deal on climate change.

This, however, raises the question why the Stern message does not translate into political action: If it is widely accepted that enhanced efforts for climate

---

[1] Throughout this book, *climate politics* refers to the international dimension of addressing climate change, that is, the forms of cooperation on climate change organised under the UNFCCC in the first place; the notion *climate governance* goes beyond these state forms of governing the climate and also includes activities of non-state actors and transnational forms of cooperation.
[2] The Intergovernmental Panel on Climate Change (IPCC) is a panel with scientists nominated by national governments; its reports regularly provide a synthesis of the available scientific knowledge on climate change.

protection are not only urgently needed but also offer economic opportunities, why are governments not willing or able to agree on an ambitious climate deal? And does this mean that the insights and arguments of the Stern Review and its broad discussion had no effect on climate governance at all?

This book claims that this is not the case. It argues that an important shift has occurred to the climate discourse in recent years of which the Stern Review is one of the most visible expressions. The increasing awareness for the economic significance of climate change, put forth by Stern and others, has led to a search for new strategies that address the economic risks and opportunities related to climate change and climate protection.

The continued failure to achieve meaningful progress in the international climate negotiations is not at odds with this hypothesis, all to the contrary. The increasingly economic understanding of climate change pushed by the Stern Review has led to a new problem framing and the emergence of government activities that seek to align climate protection with economic objectives, but give less emphasis to the role of international cooperation through a Kyoto-like treaty. This new framing of the climate challenge does not emerge from a coherent process, but rather from a whole series of developments in climate change governance in recent years that contribute to an understanding of climate change as an economic challenge.

1.1.2 The emergence of the climate investment discourse

One of these developments is the enhanced emphasis that is given to the finance and investment needs for climate related activities. To be clear: Climate change and its regulation always entailed an economic and financial dimension. However, the financial side of achieving climate protection has played a minor role in climate debates for long and was addressed rather implicitly. In the carbon market discourse, for instance, the focus has been on market efficiency rather than on the financial flows created by these markets.

This has changed remarkably in recent years. The financial side of climate climate protection has moved to the centre of climate change debates and strategies. Following the Stern Review, studies estimate the costs of climate protection and a transformation to low-carbon economies in different sectors or world regions, and consider potential financial sources. In different domains of climate governance, governments and other organisations deploy a great number of new policies and instruments to enhance investment for climate protection.

The analysis in this book focuses on three related developments that high-light the investment challenge related to climate protection. First, raising investment has become an important issue in the *climate finance field*. The provision of climate finance to developing countries turned into a central concern in the climate change negotiations after 2007. The focus of the climate finance debate also shifted, from supporting adaptation to realising emission reductions in developing countries. The much higher levels of climate finance that would be required in consequence turned the spotlight on the private sector. Governments and international organisation create policies and instruments to encourage private investment to climate related activities in developing countries. The World Bank Climate Investment Funds are the most prominent example here.

Second, raising investment also plays an important role in *strategies for low-carbon growth and development*. Governments formulate these strategies to orchestrate a transformation to climate sound economies, and to realise the economic benefits from building low-carbon industries and sectors. Organising a low-carbon transformation, according to many of these strategies, requires a more active role of the state in the economy, in particular in supporting low-carbon investment activities.

Third, the role and success of carbon market mechanisms like the Clean Development Mechanism (CDM) are increasingly assessesd against their potential contribution to climate finance and investment levels. These debates suggest that using carbon market mechanisms as a tool for climate finance would require substantial changes to the governance of these mechanisms, to allow more government control in directing financial resources to the required places and uses. Very similar concerns play an important role in the creation of the REDD mechanism for financing forest protection: one of the main struggles in designing REDD is to provide incentives for private investors while guaranteeing the environmental integrity of invesment into the forerstry sector.[3]

Taken together, these developments form what I will coin the *climate investment discourse*. Governing finance and investment for climate protection apparently has become a more important task for states and other public agencies in recent years, and seems to require new forms of intervention and regulation.

It is obvious that the new emphasis on finance and investment and the huge attention given to the Stern Review are somehow related: both address the economic side of climate change and, more specifically, highlight the financial challenges related to climate protection. Also, both issues gained importance roughly at the same time, though it is difficult to determine exactly when the climate investment discourse took of.

However, it is important to describe in more detail how these two developments are related to each other, to also understand the nature and significance of

---

[3] REDD is the acronym for Reducing Emissions from Deforestation and Degradation, and is used here as generic term for an evolving instrument, see chapter 5.3.

the changes to climate change governance in recent years. One possible way forward in that sense would be to describe the *economisation* of climate change through the Stern Review and the climate investment discourse: this analysis would highlight how climate protection strategies are increasingly assessed in economic terms and made subject to market logics, and it could build on a broad literature that addresses the extension of markets or market logics to new sectors or domains.

1.1.3 And a return of the state

However, simply describing these processes as economisation of climate governance fails for at least two reasons. *First*, the regulation of climate change has been deeply penetrated by economic perceptions, interests and calculations long before the publication of the Stern Review: the 1992 UN Framework Convention on Climate Change and the 1997 Kyoto Protocol call for and institutionalise cost-effectiveness as a primary principle in climate governance, and the creation of carbon markets has made the choice of climate protection activities subject to market decision.

An economisation perspective that focuses on the extension of markets or the imposition of market logics would therefore, *second*, miss the exact nature of the recent changes to climate governance. It is not enough to conclude that the Stern Review or the climate investment discourse strengthen an economic perspective on climate change and turn climate protection into an economic challenge; rather, it is crucial to highlight how this is different from the economic interpretations and strategies that had already played a crucial role in climate governance before these developments.

One key question in this regard is the role that is ascribed to market and state activities in climate governance, and the particular understanding of the market-state relation that follows from this. The approach to climate protection adopted with the Kyoto Protocol gives great emphasis to market solutions and limits the scope for state interventions on behalf of the climate accordingly: market solutions are expected to allow for the most cost-effective solution to climate change.

The Stern Review and the climate investment discourse seem to challenge this market narrative. The Stern Review assesses climate protection strategies according to their economic costs and benefits, but also supports enhanced state interventions to maintain global warming at an acceptable degree. The lead author of the Stern Review and others call for new forms of regulation to make markets work for climate protection and allow for a new era of sustainable

17

growth and prosperity this way (Stern 2009a). Costanza (2009: 1107) even suggests that following Stern´s proposals 'could change capitalism for the better'.

Likewise, the approach that is taken to finance and investment for climate protection ascribes greater responsibility to states in managing the climate crisis. New Public Finance Mechanisms that support private investment to climate protection aim at directing private finance flows to specific purposes, instead of leaving funding decision to market choice. Likewise, many of the current proposals for reforming the CDM would limit rather than enhance market flexibility in climate protection.

The question, however, is whether this new responsibility of the state (or public authorities more general) challenges the market-driven approach to climate protection that developed in the years after the Kyoto Protocol, and whether it gives governments more flexibility in regulation climate change, for example to realise social and environmental objectives that go beyond emission reductions (Newell and Paterson 2010).

The investment perspective on climate change is accompanied by the optimistic assumption that focusing on the (investment) opportunities related to climate change rather than the costs of climate protection could help to overcome some of the obstacles that climate governance faced for many years: The expectation in that sense is that large shares of global investment capital are not bound to emission intensive industries and sectors and could be channelled to low-carbon activities if the appropriate investment environments were in place.

One fundamental challenge for states is thus to put in place enabling frameworks for low-carbon investment. While aligning climate protection with the interests of investors that way is seen as a more realistic approach to achieving environmental objectives, it also brings a new group of powerful business actors and their particular interests to the centre of climate change governance. The analysis of the climate investment discourse thus takes a closer look at the enhanced responsibility that is ascribed to states for aligning economic growth with climate protection objectives by addressing investment conditions, and at how much freedom this allows for states in managing the climate crisis.

1.2 The research project

The book thus pursues *two related objectives*. The first objective is to explain how the Stern Review, or rather the discussion around it, reframed climate change as an economic challenge, and how the emergence of the climate investment discourse is linked to this new framing. Instead of reducing the link between these developments to a causal relation, describing the latter as a consequence of the former, this book understands both as the expression of and contribution to the same shift in the climate change discourse.

The second objective of the book is to reflect the significance of this shift for climate change governance. It argues that both the Stern Review and the climate investment discourse are part of and contribute to a particular economic perspective on the climate challenge that is distinct from what has been described as *marketisation of climate change*: The growing awareness for the huge economic risks and opportunities related to climate change and climate protection gives rise to more active government strategies in managing the climate crisis and organising a low-carbon transformation. In analysing these strategies, one central concern is therefore to identify and reflect the changing roles of markets and states in climate governance.

We can describe the objectives of the book through *two related questions* therefore:

- *Question 1*: How do the Stern Review and its discussion re-frame climate change as an economic challenge, and how does this framing effect on climate governance?

- *Question 2*: What role is ascribed to the state in addressing the economic risks and opportunities related to climate change, in particular for raising finance and investment for climate related activities? And how is this different from the market-driven approach to climate protection instituted with the Kyoto Protocol?

The *general hypothesis* regarding these questions follows from what has been said above about the Stern Review, the emergence of the climate investment discourse and the return of the state. The book claims that a *discursive shift* has taken place in climate change governance in recent years: The raising awareness for the economic significance of climate change leads to new strategies for aligning climate protection efforts with economic objectives. We can specficy this hypothesis by identifying three related changes that indicate the current direction of climate change governace:

- *From global to domestic*: With the increasing awareness for the economic significance of climate change, the primary focus of climate governance shifts from the level of international negotiations to the role of domestic policies and frameworks. This is sometimes interpreted as steps towards

19

realizing internationally agreed objectives, but sometimes also as replacing the pursuit of international agreement.

* *From threat to opportunity*: The increasing attention for the economic opportunities related to climate protection activities, together with efforts to identify and make visible these opportunities, enable new climate protection strategies that are centred around the positive story of a *low-carbon transformation*.

* *From current bad to future good*: Related to this, the emphasis in climate governance shifts from the economic acivities that contribute to the climate crisis, to the alternatives that are meant to replace these harmful activities in the future. It is in this context that *financing* climate protection evolves as a new challenge and strategy.

1.2.1 State of research: Stern and investment for climate protection

The questions raised in this book are hardly addressed in the existing literature on climate change governance. We saw that the Stern Review has played a central role in climate change discourse and governance in recent years. The study is also subject to extensive review and criticisms.

However, the wide majority of these contributions do not ask for the impact that the Stern Review has on climate governance, and even less so seek to explain this influence. The few existing exceptions highlight that the changing understanding of climate change as an economic challenge has raised the importance of the issue and made the climate discourse accessible to new actor groups (Egner 2007; Luks 2008). They do not specify the nature of this new understanding, however, and how it is different from the market-oriented approach to climate protection since the adoption of the Kyoto Protocol.

Likewise, the need for raising investment for climate protection is widely acknowledged and usually explained with the need to realise – and that is: to finance – large emission cuts quickly. What is not questioned, let alone exlained, is why the issue has been taken up so broadly all of a sudden in recent years, and neither is the particular approach taken to raising investment for climate protection and its significance for climate change governance.

The existing literature on the Stern Review and the role of finance and investment for climate protection differs considerably from the analytical objective of this book therefore. Nevertheless, these contributions play an important role for this analysis in different ways.

More than five years after its publication, the Stern results continue to be an important reference point for making the case for enhanced climate protection efforts. The Stern Review also repercussioned strongly in the academic world. It was followed by a great number of studies that enquire into the economics of climate change for different sectors or world regions, and thus confirm the relevance of the Stern perspective on climate change.

There are two other types of contributions on the Stern Review that are more immediately relevant for the analytical purposes of this book. First, the assessment of its methods and results among other economists highlights that the outcomes presented by Stern and his team do not offer radical new insights into the economics of climate change, and are thus not sufficient to explain the enormous importance that the study gained in climate change debates (Weitzmann 2007; Barker 2008; Schneider 2008; Yohe and Tol 2008). Second, a number of contributions highlight more in general the limits to cost-benefit analyses of climate change (Ackermann 2007; Spash 2007; Hulme 2007b). Taken together, these two groups of studies show that the results in the Stern Review depend heavily on assumptions and normative choices, and that the bold argument for the economic benefits of climate protection is grounded in political considerations rather than scientific rigor.

Regarding the financial side of climate change and climate governance, the social science literature largely focuses on carbon markets. Carbon market mechanisms have evolved as the centrepiece of climate change regulation since the adoption of the Kyoto Protocol, both within and beyond the global climate governance architecture under the UNFCCC. There is, accordingly, a vast social science literature on the emergence, governance and political economy of carbon markets, and on their justice and efficiency effects.

The great emphasis on the achievements and shortcomings of carbon markets may also be the reason why the analytical social science literature gives little attention to the broader perspective on finance and investment for climate protection introduce above.[4] However, the more important role of finance and investment in climate governance is addressed in many publications from governments, international organizations, business and civil society organisations, and think tanks or consultancies.

These contributions form the empirical material for the analysis of what I coin the climate investment discourse. The choice of this material is already an important first step in defining this discourse, as investment for climate protection is not a clearly defined object: It is addressed as the private side of climate

---

[4] The few existing exceptions analyse the perspective of institutional investors on a low-carbon transformation (MacLeod 2010), raise the question of democratic decision making in investment decisions related to climate change (Lohmann 2009), or develop a heuristic for explaining the patterns and direction of clean investment flows within a complex and fragmented transnational governance landscape (Newell, Jenner et al. 2009).

finance, within strategies for a low-carbon transformation, and in the context of carbon market governance.

Many of the contributions that address finance and investment for climate protection take a governance perspective and often pursue explicit political objectives. These studies: identify ways to scale-up climate related investment, including through domestic regulation and Public Finance Mechanisms; study the effectiveness of the climate finance architecture and make proposals for its reform; focus on the finance needs of developing countries and the institutional setting for delivering climate finance; and ask for the role, perspective and interests of investors.

A second type of studies analyse the political economy of climate related investment, including country-specific investment conditions and risks, the role of climate legislation, and the growth opportunities related to investment in renewable energies and clean technologies. The wider benefits of moving to less emission intensive economies are also central to strategies for low-carbon growth and development; these often focus on the political economy of energy and emission intensive sectors in particular countries and include estimates on the investment needs for transforming these sectors. Low-carbon strategies and contributions to the Green New Deal (GND) debate, finally, also emphasize the need for and the benefits from political intervention to direct investment to green sectors and technologies.

1.2.2 A discourse perspective on the economics of climate change

This book seeks to shake the common understandings of the economic and financial challenges related to climate change, by analysing both the Stern Review and the climate investment discourse from a discourse theoretical perspective. This analysis is based in poststructuralist ontology, which claims that the economy is, just as any other object or domain, discursively produced. Economic rationalities are historically and spatially distinct, and in their concrete form constituted in discursive practice (Jessop and Oosterlynck 2008).

Discourse theory and the economy

The understanding of the economy as discourse is an important issue in the writings of those authors who today are seen as the main sources for poststructuralist theorising, albeit in different ways.[5] One main starting point for the discourse theory of Ernesto Laclau and Chantal Mouffe was the remaining economic essentialism in Marxist theory, in particular in Gramsci's hegemony concept. One core objective for Laclau and Mouffe was to highlight that the position of particular classes or social groups within the economy, and their own understanding of their role and interests, are constituted in discursive processes, and subject to change accordingly (Laclau and Mouffe 1985).

Michel Foucault addresses the discursive constitution of the economy in particular in his 1978 and 1979 lectures at the Collége de France. Here, he describes how the emergence of economics as an independent scientific discipline within a new governmental rationality constitutes the economy as an autonomous domain. This was the starting point of our present understanding of the economy as a separate field of society with its own rules and logics (Foucault 2007). Foucault's 1979 lectures also identify important differences between (classical) liberal and neoliberal rationality with respect to the regulation of markets and the economic sphere.

Today, discourse theoretical or poststructuralist perspectives are applied to different economic domains in an increasing number of social science subdisciplines. One widely shared objective of these approaches is a de-essentialising perspective on the economy: they reject the idea of economic or capital logics that determine the form of political organisation or social formations. Economic relations, according to this perspective, are one principle of social organization among others and not the dominant one a priori. This, however, does not neglect the possibility that economic relations are the dominant principle in a concrete social formation, as Foucault has repeatedly stressed as well.

A poststructuralist perspective thus takes a relational approach to the economy. The historically concrete forms of the economy are 'the expression of previous historical and contingent struggles for a certain form of socio-economic organization and a specific spatio-temporal structuration' (Wullweber and Scherrer 2010: 13).

More specifically, discourse theoretical perspectives have been applied to financial markets and financial practices, including in the field of International Political Economy. De Goede (2004, 2005) analyses the emergence of the modern financial system from a poststructuralist perspective, to challenge the understanding of financial markets as bounded powerful system that develops in a necessary or evolutionary process. Instead, she highlights the historical and discursive processes 'through which a domain we now call finance has material-

---

[5] Poststructuralism is understood here as assuming radical contingency and discursively constituted meanings.

23

ized' (de Goede 2005: 7). Her analysis describes how finance discourses and finance theory have contributed to depoliticizing financial practices and portraying finance as a purely technical matter. This aspect is highly relevant in analysing the climate investment discourse as well.

A second important perspective for addressing the economy within post-structuralist ontology is the governmentality concept put forth by Foucault. Other differences notwithstanding, a governmentality perspective shares with discourse theory the assumptions of radical contingency and discursively produced meaning. The governmentality concept focuses on the relation between governmental thought or rationalities and governmental practices, and gives emphasis to their mutual constitution. It thus highlights the formation of objects and identities in discursive practices, including objects of government.

One key area of early governmentality studies was the economy and the government of economic life. A governmentality perspective does not treat the economy as a distinct sphere and analyses its relation to state or society. Instead, it addresses the economy like any other field of society, paying attention to the practices that constitute its different units: 'One could chart the problem spaces within which these zones had been delineated and brought into existence as calculable and manageable spaces – the national economy, the industry, the factory. One could examine the emergence of the forms of knowledge and expertise such as that of management, human resources, or accountancy, which tried to render these spaces thinkable and develop tactics to govern them' (Rose, O'Malley et al. 2006: 94).[6]

A governmentality approach as a de-essentialising perspective on the economy emphasizes that there is no singular economic logic or rationality. Rather, the understanding of what is economically rational is historically distinct and depends on forms of knowledge and (intellectual) practices. From a governmentality perspective, 'those gray and tedious sciences of economics, management, and accounting could be seen once again – as they had been by Marx, Weber, Sombart, and many other theorists of capitalism – as crucial for making up and governing a capitalist economy' (ibid. 95, Power 2000; Miller 2008).

---

[6] In their focus on the practices that constitute the economy (or other domains of society), governmentality studies are related to other sub-disciplines in the social sciences that describe the relational character of markets and the economy. Cultural Economy and historic sociology study the creation of markets as social processes and thus pay attention to the 'performativity of economics' (MacKenzie and Millo 2003: 107); economics, in that sense, is itself 'a form of representation and technological practice that constitutes the spaces within which economic actions is formatted and framed' (du Gay and Pryke 2002: 2). Describing markets as fields that are shaped by the activities of participants and thus the power relations between them, several contributions to the sociology of markets highlight that markets are an expression of cultural practices (Fligstein 2001). Actor Network Theory adds another promising perspective for studying these processes.

A poststructuralist perspective emphasises that (scientific) analysis of the economy is always a form of articulation, and thus the creation of something new. A recent edited volume with discourse analytical contributions to the economy distinguishes between the discursive constitution of the economy and of economics as a scientific domain (Diaz-Bone and Krell 2009). The contributions on the latter emphasise that economic analysis, rather than a mere representation of economic activities, is an active intervention and contributes to the constitution of economic spheres and objects. Deconstructing these practices is crucial to understand what counts as economic, and where and how the boarders between economic and non-economic domains are drawn (Stäheli 2008a).

The constitution of climate change as an economic challenge

The analysis of climate change in economic terms is thus not the application of predefined perspectives and understandings to a new issue area or problem, but constitutes climate change as an economic challenge in a particular way. The first objective of this book, in that sense, can be specified as identifying the way in which climate change has been (re-)framed as an economic challenge in recent years, the particular economic understandings that are used and produced in this process, and how this in turn allows for a greater role of economic perspectives, actors and practices in addressing the climate challenge.

The second and related objective of the book is to explain how the need for addressing investment for climate protection has emerged within this particular economic understanding, how the investment challenge is framed accordingly, and if and how the climate investment discourse changes the governance of climate change. As said above, one crucial issue here is the changing understanding of the market-state relation in climate governance.

This is not meant to claim that a discourse perspective can replace the studies on investment needs and opportunities related to climate change –these are important for formulating political strategies. And neither does a poststructuralist perspective, which asks for the discursive constitution of investment as an object of government, neglect the existence of structural forces that make it necessary to address investment for climate protection in a particular way.

All to the contrary, structures are at the very heart of poststructuralist ontology and its primary object of analysis. However, a poststructuralist perspective highlights that our understanding of structures are discursively produced. Structures have constraining and enabling effects only through discourse, and these effects are always historically specific therefore: A poststructuralist perspective is a way of understanding how the concrete form of framing the investment challenge enables certain forms of political intervention while excluding others.

Looking at the development of climate protection strategies this way means to pay attention to the role of power in governing climate change, if power is understood as the mechanisms that temporarily fix the production of meaning, and thus create the context and define the way in which objects of government can be addressed (Moebius and Reckwitz 2008). Tackling climate change successfully, Newell and Patterson (2010: 188) suggest, requires 'bringing together people who could never have previously imagined working together – environmentalists with venture capitalists'. Reflecting the new perspective on the climate challenge that evolves from these constellations is the main analytical purpose of this book.

1.2.3 Governmentality and climate change

Building on Foucaultian terminology, the book deploys a governmentality perspective to highlight the mutual constitution of rationalities and practices in addressing climate change as an economic and investment challenge. I will first give a short overview of how the governmentality perspective has been used to analyse global climate governance, to then highlight how my approach is different from most of these studies.

Governmentality studies on climate change

A governmentality perspective has been used to analyse the rationalities, technologies and subjectivities involved in governing different fields and domains of climate governance. What these studies have in common is that they highlight the process of constituting political problems and solutions, that is, of rendering objects governable.

It is possible to distinguishing three types of governmentality studies on climate change (Rothe 2011). First are those studies that introduced the governmentality perspective to the field of climate politics and can be seen as pioneering studies in that regard (Oels 2005; Bäckstrand and Lövbrand 2006). These studies demonstrate the techniques and rationalities that make climate change governable and structure the way we act upon it.

More recently, governmentality studies on climate change focus on the (micro-) practices of governing different domains of climate governance, offering in-depth analyses of carbon offsetting technologies (Lovell and Liverman 2010), carbon accounting practices (Lövbrand and Stripple 2011), the technologies of governing individual's carbon emissions (Paterson and Stripple 2010), or the epistemological frameworks involved in global climate science (Lövbrand, Stripple et al. 2009). Rather than asking for the governmental rationalities that structure the activities in certain domains of climate governance, these studies take a bottom- up perspective and investigate the specifics of the practices involved in governing the climate.

Finally, the governmentality perspective is increasingly combined with other theories and approaches that share a discourse theoretical or poststructuralist ontology, such as actor-network theory (Blok 2010), hegemony theory (Methmann and Rothe 2011), or the concept of securitization (Oels 2012).

Moreover, governmentality studies have enabled important insights into the government of migration as part of the global climate regime, the performativity of the annual climate change conferences in shaping the conduct of those who attend these conferences (Death 2011), and the writing practices of the Intergovernmental Panel on Climate Change (IPCC) (Hughes 2011). Reflecting the growing importance of forests in the global climate regime, governmentality studies have also highlighted the technologies and subjectivities in governing forests and forest carbon stocks as part of climate protection (Stephan 2012).

Following the increasing awareness within discourse theory and governmentality studies for the need to study the politics of identity, governmentality studies concerned with climate change also address the identities and processes of identity formation of climate change activists (Dowling 2010), and the processes of making individuals into agents of reducing carbon emissions (Paterson and Stripple 2010). Finally, a governmentality perspective is also increasingly used to examine the technologies of government involved in climate change adaptation programmes (Keskitalo, Juhola et al. 2012).

Analysing change to the rationality of climate governance

Governmentality studies thus make important contributions to understanding how climate change is made visible and governable in different arenas. On the one hand, this book follows the same objective: it aims to show how investment for climate protection is rendered visible and governable more recently.

On the other hand, this book goes beyond describing the processes of making objects visible and governable as well, and is different from a great part of the governmentality literature on climate change therefore. As Rothe (2011: 7) points out, most of these studies 'restrict themselves to an understanding of climate governance, i.e. they basically ask how climate change is governed. There

27

is no article that asks why climate change in a certain regime of practices is governed in a specific way and not in another'.

Following this lead, we can say that the analysis in this book does not restrict itself to describing the processes that constitute governmental objects in the domain of climate change, but also asks for the significance of these processes: It seeks to understand why these objects are made visible and governable in the way they are, and how this can be explained within the general rationality of governing climate change and changes to it.

Such endeavour entails two fundamental challenges. First, this approach claims the existence of such rationality. In a certain way, this is at odds with more recent developments in the study of governmentality, which emphasise the need to pay attention to the existence of different and sometimes competing techniques and rationalities of government. Recent governmentality studies on climate change, accordingly, focus on specific practices of government within their respective locales and contexts in many cases (e.g. Lovell and Liverman 2010). At least partly, this is a reaction to the critique that governmentality studies assign practices to predefined logics such as neoliberalism or advanced liberal government, and thus lose sight of what is specific about the respective practices.

However, this book does not claim that it is possible to identify a certain logic or rationality that determines governmental practices. All to the contrary, the theory section in chapter 2 will introduce an understanding of governmentality that highlights the mutual constitutions of practices and rationalities. Building on the dispositif concept, it develops a research strategy that allows understanding rationalities as the common ground of a set of otherwise diverse practices. It is thus possible to identify changes to the rationality of governing climate change, without claiming that this rationality determines the practices in the field of climate change.

The second important challenge for a research strategy that seeks to analyse a change to the rationality of governing climate is to identify and delineate the rationality in climate governance prior to these changes. It is here where I draw on the studies that have been described as pioneering studies in introducing the governmentality perspective to the study of global climate politics. They describe the ways in which the global environment (and more specifically: climate change) is made visible and governable, and highlight the increasing importance of economic thought in governing global environmental processes.

This literature largely builds on the concepts of *green governmentality* and *advanced liberal government(ality)*. Luke (1996) introduced the *green governmentality* concept to describe the role of environmental studies in bringing about a specific power/knowledge formation that makes visible and manageable global environmental problems, and thus help extending government control – or bio-power – to the entire planet. The same perspective has been applied to the role of climate science (Bäckstrand and Lövbrand 2006).

Luke's understanding of green governmentality links the management of the global environment to the maintenance and expansion of capitalist growth. The new eco-knowledges aim at inserting ever larger parts of the natural world into the 'machinery of global production' and that way are the precondition for 'diffuse projects of ecological modernization' (Luke 1996: 3: 105, cf. Oels.).

More recent studies that focus on climate change governance suggest instead to distinguish green governmentality from ecological modernisation strategies: the latter, they claim, are embedded within neoliberal or advanced liberal governmentality and foreground the economic dimension of environmental or climate protection strategies. In its weak version, ecological modernisation is a technocratic economic discourse that advocates free market settings and limited government interventions as a way of incentivising technological innovation and achieving cost-effectiveness in environmental protection.

In climate governance, the weak ecological modernisation discourse found its clearest expression in the Kyoto Protocol and the importance that carbon markets gained for achieving cost-effectiveness in climate protection. It is against this understanding of *market-driven climate governance* that this book identifies and interprets changes to the rationality of governing climate change. It claims that the Stern Review and the climate investment discourse question the dominance of the market superiority narrative: both ascribe enhanced responsibility to the state in achieving climate protection and a low-carbon transformation.

It has already been said that the discourse theoretical and governmentality literature on climate change does not address these changes. The studies that analyse the rationality of governing climate change usually do so prior to the publication of the Stern Review. More recent studies address the specific ways of making carbon tradable or extending carbon markets to new areas of climate change governance, but pay little attention to the changing role and governance of carbon markets in recent years, and what this implies for the rationality of governing climate change (MacKenzie 2009; Lovell and Liverman 2010; Lansing 2011; Lövbrand and Stripple 2011). Likewise, the new or changing role that is ascribed to the state in governing climate change, and in enabling and raising investment for climate protection, is a not a topic in the social science literature on climate change thus far (one exception in that regard is: MacNeil and Paterson 2012).

1.2.4 Operationalisation, research strategy, and structure of the book

The book takes this gap in the analysis of climate governance at its starting point: It asks how the growing awareness for the economic risks and opportunities related to climate change (as expressed in the Stern Review) changes the approach to governing climate change, and describes the new emphasis on investment for climate protection activities as one possible form of reacting to this awareness. The ultimate objective is to question the specific way in which climate change is framed as an economic and investment challenge, to highlight the opportunities and constraints that result from this approach, and to make visible alternative approaches to financing climate protection that are marginalised in current discourses.

As indicated above, the research is theoretically grounded mainly in the work of Michel Foucault and writings in critical discourse theory. This involves a poststructuralist ontology, that is, to put it very briefly, the notion of *discursively produced meaning* and *radical contingency*: Social relations and structures are seen to be constitutively incomplete, and this *structural undecidability* (Derrida) gives room to political practices and change to discourses.

The analysis is organised by research a strategy that Howarth and Glynos (2007) present as articulation. Opposed to deductive or inductive approaches, this problem driven approach does not aim at theory testing through case studies, or theory generation through abstraction. Rather than 'proving' a causal relation, the concept of articulation aims at a convincing and in itself coherent explanation. It is thus a way of specifying a poststructuralist research strategy that highlights the relationship between social structures, political agency and power.

Howarth and Glynos suggest to operationalise this analytical perspective through the construction of a single explanatory chain. It consists of the *problematisation* of empirical phenomena that require explanation; the *retroductive explanation* of these phenomena; their *interpretation* in the light of existing theoretical and explanatory concepts; and *critique* in form of 'deconstructive genealogy' that discloses possible alternative framings (ibid. 155).[7]

This book therefore constructs a single explanatory for analysing the Stern Review and the climate investment discourse, using different theoretical concepts for the individual steps of analysis. After introducing and developing its analytical concepts in chapter 2, the book proceeds in three main steps.

Chapter 3 outlines and explains the economic turn in climate governance. It treats the Stern Review as a problematisation that establishes a certain understanding of climate change as an economic challenge. Against this background,

---

[7] The methodological issues and challenges related to such an approach will be taken up again and specified in chapter 2, after developing the theoretical instruments for this book.

the second step analyses the climate investment discourse as a particular form of responding to this challenge: *Chapter 4* describes the constitution of investment as an object of government, and *chapter 5* examines instruments and practices that address the investment challenge in climate governance. *Chapter 6* as the third step brings together the results of the empirical analysis and asks whether and how the investment turn entails a change to the rationality of governing climate change, focusing on the new role that is ascribed to the state in climate governance more recently. The *conclusion* in chapter 7 reflects the analytical approach of this book, and returns to the question raised in the beginning: What the economic problematisation of climate change suggests for the future of international climate politics.

2 Theory: Power/Knowledge

'It is not a matter of emancipating truth from every system of power (which would be a chimera, for truth is already power) but of detaching the power of truth from the forms of hegemony, social, economic and cultural, with which it operates at the present time'

(Foucault 1980b: 133).

After more than two decades of reception and discussion of his writings, it can be said that, throughout his work, Michel Foucault fundamentally addressed one question: how formations of knowledge have become truth and the basis for political activity. The objective of his work, he suggests, 'is to see how men govern (themselves and others) by the production of truth' (Foucault 1991: 79). This *history of thought* is the central theme that runs through Foucault's entire work: In the beginning, he aims at a purely discourse theoretical account; later, he opens this perspective, to analyse the mutual constitution of power and knowledge.

The double notion of *problematisation/rationality* is crucial for understanding this relation: Foucault claims that events, practices, power relations and knowledge formations are intrinsically linked. In a process of problematisation, they constitute a particular rationality, which enables and delimits political action. Problematisation thus describes the formation of domains of acts and thoughts that become the object of governmental practice.

The notion of problematisation appears in the last phase of his work only but is, as he then explains, the basic concept that defines the common ground of his studies, the 'one element that was capable of describing the history of thought' (Foucault 2005b). Foucault's empirical studies do not aim at assessing particular political practices, but are concerned with the question of how an object of government is defined and made governable.

The analysis of problematisations therefore assesses political practices not on grounds of the understandings that prevail in political discourses, but makes the formation of these understandings a part of the analysis. The history of thought is thus not identical with the history of ideas or ideologies and does not take these historical positivities as its object of investigation, but addresses the entire field in which thinking as an activity happens.

33

The analysis of problematisations thus describes the field of constraints and enabling conditions in which political practice takes place (Kerchner 2006). This analysis entails 'the historical, yet *a priori* conditions that make thought and practice possible and that, as such, are properly said to govern them both' (Thompson 2010: 127).

Problematisation as a methodology therefore combines and rearranges Foucault's two crucial methods, archaeology and genealogy (Lemke 1997). Archaeology is used 'to examine the forms of problematisation themselves' (Foucault 1986: 11-12), it describes the form in which human beings understand and describe themselves and the contexts they live in. Genealogy accounts for the formation of problematisations 'out of the practices and their modifications' (Foucault 1986: 11-12). It describes the historically specific structures and conditions through which a matter is put at issue and defined as a political problem.

The notions of problematisation and rationality define the analytical perspective of this book. It asks how the importance of the Stern Review is related to the emergence and significance of governmental practices that address investment for climate protection, and how these developments contribute to and reflect changes in the rationality of governing climate change.

The book is a study of governmentality therefore. Foucault introduced the governmentality concept in his Paris lectures in 1978 and 1979, to highlight the mutual constitution of governmental rationalities and practices: Governmental activities constitute the objects they address within a certain rationality of government and, vice versa, modificate governmental rationality. The governmentality concept also reflects Foucault's understanding of power as a productive force, and is the context in which Foucault addresses the genealogy of the state.

The governmentality perspective is crucial for understanding how objects or domains are governed within a particular power-knowledge formation therefore. However, there are also two problems related to the analysis of governmentality. The first problem is the unclear status of governmentality as either ahistorical concept or a historically specific category (see section 2.2). This book therefore makes use of the insights from the governmentality concept, without using the concept itself.

The second problem related to the study of governmentality is a critical issue Foucault's work more general: How to explain the emergence and change of coordinated forms of power. Foucault's studies focus on 'the coherence and systemacity' (Collier 2009: 94) of a discourse or governmental constellation: they indicate that something has changed compared to a previous formation, but cannot explain, the transition between those systems.

Instead of building on alternative theoretical approaches to fill this gap, this book argues that it is possible to address the issues of change and coordination through the notions of dispositif and problematisation. The dispositif as *analytical grid* describes how discursive and non-discursive elements within a certain power-knowledge formation are linked and directed towards a strategic objective, and allows decoupling the emergence of rationalities from the interests of individual actors and the power of a hegemonic force. While this seems leave little room for agency, the notion of problematisation gives emphasis to the role of thought in translating historical challenges into governmental rationality.

The theory chapter therefore proceeds as follows. The first part introduces Foucault's main analytical methods and concepts. The second part introduces the governmentality perspective, and describes how it is used for the analysis in this book through the concepts of dispositif and regimes of practices. The third part explains the operationalisation of this research perspective, and introduces the *single explanatory chain* as a poststructuralist methodology for the analysis of the Stern Review and the climate investment discourse.

2.1 Epistemology: Archaeology, Genealogy, Power

Archaeology and genealogy are the two fundamental methods that Foucault uses in his empirical studies. Whereas the latter replaces the former when he turns to the question of power, insights from archaeology remain important for the study of power-knowledge complexes and are entailed in the concepts governmentality and problematisation as well.

2.1.1 Archaeology of discourse

The first, archaeological phase of Foucault's work culminates in the *Archaeology of Knowledge* (AK) (Foucault 1989b). In this largely methodological book he synthesises the discourse theoretical method he deployed in his preceding studies on madness, the birth of the clinic and the human sciences. At this time, Foucault does not aim at a general theory of the formation of discourses, but focuses on the rules of the formation of expert discourses; he is only interested in 'what experts say when they speak as experts' (Dreyfus and Rabinow 1982: xxiv).

Describing in short the objective, central argument and main categories of the AK is crucial, on the one hand, to clarify important issues regarding Foucault's epistemology, in particular his understanding of discourse and knowledge; and on the other hand, to understand why and how he abandons a purely discursive account and turns to genealogy.

What Foucault's first studies had in common was the question how a certain way of thought is established as historical truth at a particular time. Beyond this, however, these studies differ in their conceptual and methodological approaches.

While *Madness and Civilization* studies the interplay of discourse and institutions in targeting and creating madness as an object of government, his next book on *The Birth of the Clinic* signifies an 'extreme swing towards structuralism' (Dreyfus and Rabinow 1982: 15), identifying the structures that sustain medical practice and discourse. But though he was 'unable to avoid [...] frequent recourse to structural analysis' here (Foucault 1989b: 16), his objective was to identify *historical conditions of possibility* rather than *atemporal* structures. 'Even at this point', Dreyfus and Rabinow (1982: 15) conclude, 'Foucault was never quite a structuralist'.

While these two books present mixed pictures methodologically as they argue both on the level of discourses and the level of institutions, his next book then comes closest to the methodology of the AK: in *The Order of Things* (OT), Foucault aims at no less than explaining the emergence and development of western thought in form of an archaeology of the human sciences. Foucault claims that this study 'does not belong to the history of ideas or of science: it is rather an inquiry whose aim is to rediscover on what basis knowledge and theory became possible; within what space of order knowledge was constituted; on the basis of what historical *a priori* [...] ideas could appear, sciences be established, experience be reflected in philosophies, rationalities be formed' (Foucault 1973: xxi, xxii).

Foucault describes this historical *a priori* as the episteme or historical order that governs the production of knowledge within a certain epoch, separating sciences from the non-scientific. The OT demonstrates that changes in the (underlying) episteme cause changes within different discourses, but remains at a purely descriptive level in not asking *how* or *why* changes to the episteme occur (Dreyfus and Rabinow 1982).

Crucial for his decision to develop an approach to archaeology that focuses on the formation of discourses and (momentarily) neglects social institutions is a fundamental epistemic shift in modern western thought that Foucault describes as the emergence of man as subject and object of knowledge: 'Man appears in his ambitious position as an object of knowledge and a subject that knows: enslaved sovereign, observed spectator [...]' (Foucault 1973: 312).

The result is that the human sciences are trapped in an analytics of finitude as man appears doubled: as a fact among other facts that can be studied empirically, and as the transcendental condition of possibility of all knowledge: 'the limits of knowledge provide a positive foundation for the possibility of knowing' (Foucault 1973: 317, Honneth and Saar 2008). Knowledge claims are thus always 'twisted' and 'warped', as 'each new attempt will have to claim an identity and difference between finitude as limitation and finitude as source of all facts' (Dreyfus and Rabinow 1982: 31).

It is against the background that Foucault decides to develop an approach to the analysis of discourses that rejects both immanence and transcendence.[8] On the one hand, he aims at a method of analysis that does not rely on a founding human subject as the origin of discourse or writing history and thus is 'purged of all anthropologism' (Foucault 1989b: 16). The objective is to study western societies just as ethnology approaches unknown cultures, without relying on the interpretations of its members (Foucault 1973). On the other hand, he rejects an understanding of discourse as representing a (materialist) real or other: Discourse has its own materiality and constitutes its objects rather than merely representing them.

Foucault therefore develops a methodology for the description of discourse formations that brackets all meaningful categories, such as tradition, evolution, or oeuvre. Once these categories are removed, the remaining multiplicity of discursive elements could be analysed independent from its hermeneutical context. The objective is not to ask for the background or real meaning of speech acts, but for their mode of existence, 'what it means for them to have appeared when and where they did – and not others' (Foucault 1989b: 109).

The objective of the AK is to explain the 'law of rarity' of discourse (Foucault 1989b: 134). The rules that govern the rarity of statements are what Foucault calls the discursive field or the law of difference between the limited number of discursive events and all the other things that could have been said (Howarth 2000). To find an explanation that is irreducible to interpretations and formalizations, and thus neither derived from hermeneutics nor structuralism, the AK seeks to find the rules of formation within the discursive formation itself, describing the surface of a discourse or its positivity (Howarth 2000, Lemke 1997).[9]

---

[8] Interpretations vary as to whether Foucault's objective with the AK was to rewrite the history of man in an account that goes 'beyond structuralism and hermeneutics' (Dreyfus and Rabinow 1982), or to develop a general approach to the study of historical discourses (Howarth 2000).

[9] Identifying the *rules of formation of discourse* that distinguish the rare discursive events from all other utterances is thus clearly different from linguistic theory or language analysis that ask for the rules according to which a statement is made.

What remains fundamentally unclear in the AK, however, is the status of the rules of formation as principle of explanation. Foucault sometimes describes these rules as purely descriptive relations; in other occasions, he speaks of *prescriptions* in the sense of a causal relation, suggesting 'that one can define the general set of rules that govern the status of these statements' (Foucault 1989b: 115, Howarth 2000).

The strong causality implied in this formulation is at odds with the very objective of the AK, to remain on a purely descriptive level. It is this tension that points to the most fundamental problem of the AK, to clarify the relation between discourses and the non-discursive sphere: The description of rules of formation emphasises that discourses do not evolve isolated from the world that surrounds them, and that the conditions for the emergence of new discursive objects originate in non-discursive relations.

Rejecting an understanding of the discursive as representation of a non-discursive *real* or *material*, Foucault describes discursive practices as dependent on non-discursive practices, but nevertheless gives priority to the former as they organise the non-discursive relations. The clinical discourse, for instance, organises medical institutions, cognitive capacities, and the positions from which medical experts speak.

It is difficult to see, however, how discourse presides over other social relations, without relying on the cognitive capacities of subjects who give meaning to discursive practices, and thus how to distinguish *statements* from other, non-meaningful utterances (Honneth 1984). The only way to uphold the simultaneous rejection of objective laws and subjective creation of meaning is a circular argument, which understands the observable and describable regularity of discursive formations at the same time as their conditions of existence.

Foucault thus abandons his own methodological master plan for a quasi-structuralist explanation of discursive formations in form of the *law of rarity*. Suggesting that it is the position within discourse that attaches value and meaning to the discursive elements and produces the regularity of discourse, Foucault leans towards a structuralist interpretation again: the rules of formation are the law that governs the scarcity of statements (Collier 2009).

It is the double objective to go beyond hermeneutics and structuralism that explains for the complex and, finally, failing method that Foucault develops in the AK. What is missing is a 'theory of articulatory practice' that defines the relation between the rules of formation of discourses and external (causal) relations that effect on these rules (Howarth 2000: 65f.).

Likewise, it lacks an explanation of discursive change: Foucault's empirical projects on madness, delinquency and the birth of the clinic all emphasize the emergence of new discursive formations; archaeology, however, focuses on the inner coherence of discourse systems and lacks a dynamic perspective on the transition from one system to another (Suárez Müller 2004). While change is understood as the intersection of two structural systems, the passage between the systems appears as a sudden shift rather than a describable transition.

These problems prompt Foucault to open up his project towards the 'question of power' (Foucault 1973: 175, see Lemke 1997: 50ff.), that is, to ask for the formation of the rules of formation outside the discursive sphere. The shift towards a form of analysis beyond archaeology is also based in Foucault's growing interest in political discourse and its relation with scientific discourses, which makes the limits to an archaeological method become more obvious (Howarth 2000).

The turn to genealogy does not mean, however, that Foucault abandons archaeology as a method altogether. All to the contrary, 'the emphases on immanence, on coherence, and most centrally on discursive conditions of possibility, are retained' (Collier 2009: 94). The concepts of problematisation and governmentality, in particular, maintain important insights from archaeology for the analysis of political discourse and formations.

Three aspects are in particular important. First, the discursive production of objects instead of the formation of discourses around existing objects: the importance of language or concepts in making objects amenable to political intervention is central to the governmentality perspective (Rose, O'Malley et al. 2006); second, the constitution of subjects – or subject positions – within discourse;[10] and third, a perspective on science and ideology that does not neglect scientific truth altogether, but aims at making explicit the – historically contingent – rules that qualify a discourse as science, and thus can account for the mutual affection of political and scientific discourses: scientific knowledge is neither the truth basis for political discourses nor an instrument of a particular group or class (Howarth 2000).

2.1.2 Genealogy, the question of power, and power-knowledge

What Foucault abandons after the AK, then, is not archaeology as a method, but the attempt to develop a theory of rule-governed systems of discursive practices that contain the rules of formation in themselves. Archaeology is complemented

---

[10] Though Lemke (1997) criticises that it is in particular this aspect that remains unspecific within the AK.

39

by a – Nietzschean – genealogical approach that moves to the centre of Foucault's thought. It focuses on the emergence and becoming powerful of discourses, rather than on their inner constitution and coherence.

Towards genealogy

In the text *Nietzsche, Genealogy, History* that marks the beginning of this methodological shift, Foucault outlines his understanding of genealogy, drawing extensively on Nietzsche (Foucault 1977). This genealogical approach is directed against a hermeneutical-interpretative understanding of history. Following Nietzsche, Foucault rejects a supra-historical or teleological perspective on history 'that always encourages subjective recognitions and attributes a form of reconciliation to all the displacements of the past; a history whose perspective on all that precedes it implies the end of time, a completed development' (152).

Nietzsche countered what he called historical 'Egyptianism' with the notion of historical *spirit* or *sense*, which 'can evade metaphysics and become a privileged instrument of genealogy if it refuses the certainty of absolutes' (152f.). The challenge for such a genealogical perspective as a 'dissociating view' is to make the discontinuity and accidental character of history visible, instead of assuming a form of linear development.

Foucault follows Nietzsche in contrasting the concept of Ursprung/origin with Herkunft/descent and Entstehung/emergence.[11] Whereas origin seeks to capture 'the exact essence of things, their purest possibilities, and their carefully protected identities' (142), genealogy distrusts the metaphysical faith in an original identity of things, claiming 'that there is something altogether different behind things: [...] the secret that they have no essence or that their essence was fabricated in a piecemeal fashion from alien forms' (142).

The genealogist therefore searches for the *descent* of things and their emergence. Instead of looking for the exclusive generic characteristics of an object, the objective is to rediscover the 'myriad events' that contribute to its emergence, to find 'the accidents, the minute deviations [...] the errors, the false appraisals and the faulty calculations' (146) that led to our current understandings.

Our current understanding and concepts are not the result of a continuous evolving of functions or purposes, but rather 'the current episode in a series of subjugations' (ibid.). Interpretation, consequently, does not – or not exclusively – take place in the historian's perspective on history, but is rather the process of

---

[11] While Ursprung/origin is used synonymous with Herkunft/descent in some of Nietzsche's texts, such as the *Genealogy of Morals*, the different use of the concepts in other writings such as *Human, All too Human* points to the different modes of understanding of the history of objects and thinking.

making history itself through defining concepts: the development of humanity 'is a series of interpretations' (151f.). Interpretations, however, that have become established as historical truths: 'Truth is undoubtedly the sort of error that cannot be refuted because it was hardened into an unalterable form in the long baking process of history' (144).

It is obvious that Foucault continues to ascribe a central role to the formation of meaning in the constitution of societies. However, he now takes into account the role of power and struggles in the emergence of always historical forms of meaning (Ewald 1978). Entstehung/emergence takes place within a context of power relations and is produced through the interaction or clash of forces.

It is the role of the genealogist, therefore, to record the history of these struggles as the history of morals, ideals and metaphysics, and to replace the 'anticipatory power of meaning' with the 'hazardous play of dominations' (ibid.). This brings us back to the relation of ideology and science: Foucault's objective is not to deconstruct a certain truth as ideology and replace it with another truth, but to describe changes to the – political, economic and institutional – system of truth production. Genealogy asks how a current, dominant perspective emerged and how others have been neglected or not realised, and is meant to explain the emergence of the historical structures that are described by archaeology (Suárez Müller 2004).

Genealogy highlights, through the notions of descent and emergence, that discourse is based in the extra-discursive, but that these extra-discursive events or contexts influence discourse only to a certain extent. 'Discourse is underdetermined by the things of which it speaks, and by the people who wield it, and even by a combination of these two. Discourse is largely determined by these two factors, but it does also have a strength of its own' (Kelly 2009: 22). It is the rules of discourse, still, that decide the way in which language relates to things.

Starting from a contemporary problem that has been diagnosed and defined within our contemporary understandings as well, 'genealogy seeks to provide a more plausible narrative of historical processes by viewing them from their proper perspective' (Howarth 2000: 76/7). This is not about denying historical facts, but about (re-)interpreting and (re-)contextualising these facts, and aims at a reappraisal of – thus far unquestioned – historical facts.

Genealogy combines two analytical perspectives in consequence: besides a diachronic form of analysis that explains the transition and change of systems of meaning, it also entails a synchronic dimension of analysis, which describes the always specific rationality or strategy of a particular system of power (Suárez Müller 2004). This points to the close relationship between institutions, discourses and practices, and Foucault identifies as the major challenge to his further work to define this power-knowledge complex.

The co-constitution of power and knowledge

Foucault deploys this genealogical perspective to develop, discuss and – partly – reject different concepts of power. In continuity with his archaeological approach he is less interested in judging the legitimacy of different forms of power than in explaining the emergence of complex knowledge-power edifices. Likewise, he does not aim at a theory of power, but rather at an analytics that describes the functioning of power in its positivity and in different concrete forms (Ewald 1978).

The mutual constitution of knowledge and (relations of) power is central to all these concepts. Genealogy is therefore on the one hand a turn to power, but on the other hand directly relevant to Foucault's understanding of knowledge and discourse as well, and thus crucial for his epistemology. Methodologically, archaeology and genealogy work together: 'As a technique, archaeology serves genealogy. [...] it serves to distance and defamiliarise the serious discourse of the human sciences. This, in turn, enables Foucault to raise the genealogical questions: How are these discourses used? What role do they play in society?' (Dreyfus and Rabinow 1982: xxv).

A first step towards describing the relation of knowledge and power is his inaugural lecture in 1970 at the Collège de France on the *Order of Discourse* (OD)(). Here, Foucault describes various mechanisms – or forms of power – that control and limit the free floating of discourse. The OD thus opens the perspective towards the social, political and economic relations that affect or define the rules of formations of discourses. The purely negative role of power, however, which is reflected in the dichotomies legitimate-illegitimate, reasonable-unreasonable and true-false, makes the OD a transitional text in defining the complex relation of power and knowledge (Foucault 1978c; Lemke 1997).

In the following years, Foucault seeks to develop a more positive or productive understanding of this relation that does not confine power to constraining knowledge. More in general, he aims at a concept of power that is not expressed in terms of the law and does not have its origin in the institutions of the state.

The objective to develop a non-juridical understanding of power derives from his empirical studies in the prison, as he found that existing conceptions of power are not useful in describing the processes he observed here[12]: 'The case of the penal system convinced me that the question of power needed to be formulated not so much in terms of justice as in those of technology' (PK 1980: 184).

In *Discipline and Punish* (DP) (Foucault 1976), Foucault thus describes *disciplines* as techniques of power in institutions like prisons, schools or monasteries, which target the individual's behaviour and body. Disciplines as a microphysics of power are the first step away from a purely negative or repressive understanding of power; they have productive effects in forming the individual and enhancing its capacities or productivity.

In the first volume of *The History of Sexuality, The Will to Knowledge* (WK), he complements disciplines with the notion of *biopower* (Foucault 1998). Distinct from today's more specific connotations, biopower refers to techniques of power directed towards the life (bios), that is, the processes that are directly relevant to the survival and wellbeing of the population, such as health and hygiene.

Though they differ in the object of government they target, disciplines and biopower are techniques or forms of power that depend on and contribute to the formation of knowledge. The co-constitution of knowledge and power is often ignored in the case of disciplines, though Foucault emphasizes that their formation must be understood as a double process that allows for new forms of knowledge and intervention, such as clinical medicine or psychiatry (ibid.).

This co-constitution is more obvious in the case of biopower. As a governmental practice, biopower addresses society as a whole and depends on collecting and systematising knowledge of the population. This statistical knowledge allows to address the performance and productivity of society as a collective body, and vice versa, these practices enhance the knowledge about society (Foucault 2007).

The notion of biopower demonstrates the interplay of the constitution of objects, the development of knowledge on these objects, and the emergence of technologies of power that address these objects. The governmental practices and techniques of biopower 'constitute the epistemological grammar of humanism, yet we must also note that the development of humanist knowledges secures the spread of bio-power' (Owen 1995: 494). Foucault does not suggest a linear process here, however, in which the ruling classes use the apparatuses of knowledge production to analyse the characteristics of populations as a basis for intervention; rather, a 'constant interplay between techniques of power and their object

---

[12] Kelly (2009) highlights that Foucault narrowly focused on French social science and political philosophy in his judgment and was ignorant of other contributions that had taken steps in the same direction, in particular the work of Max Weber and the Frankfurt School of Critical Theory.

gradually carves out in reality, as a field of reality, population and its specific phenomena' (Foucault 2007: 79).

In *Discipline and Punish*, Foucault thus concludes that power and knowledge directly imply one another, 'there is no power relation without the correlative constitution of a field of knowledge, nor any knowledge that does not presuppose and constitute at the same time power relations' (cf. Howarth 2000: 77). Truth has a history and is produced in the formation of both objects and subjects; producing truth does not mean to invent true statements, but to establish domains that allow for the practice of distinguishing true and false.

2.1.3   Government as power

Beyond a new perspective on the power-knowledge nexus, Foucault's turn to genealogy entails a more general examination of the nature of power. He introduces and develops two related but nevertheless distinct perspectives on power in modern societies. The notion *microphysics of power,* on the one hand, describes the ubiquity of power relations and a constant clash of forces between individuals. Biopower, on the other hand, is refined toward the more encompassing concept of *government* or *governmentality,* to describe the organised ways of acting on the behaviour of individuals and society.

While it is sometimes claimed that the latter understanding has replaced the former, it is crucial to see how Foucault combines both perspectives in systematising his understanding of power. Ultimately, he distinguishes games between liberties, strategic forms of government, and moments of domination.

A productive understanding of power

Disciplines and biopower emphasise the positive – or non-repressive – and productive character of power. It is in the WK that Foucault makes the understanding of power as a productive force most explicit, criticising the common juridical understanding that describes power as the execution of the law: 'It is a power that only has the force of the negative on its side, a power to say no; in no condition to produce, capable only of posting limits, it is basically anti-energy; finally, it is a power whose model is essentially juridical, centred on nothing more than the statement of the law and the operation of taboos' (Foucault 1998: 85).

Against this, he describes (in DP and WK) several features of a positive understanding of power (Kelly 2009: 38f.): first, power is *impersonal* in that it is not held by an individual or group, but rather describes a relation between these individuals or groups; power is thus, second, *relational*; third, power is *non-centred*, as it is not held by a group or class nor has its origin exclusively in one place, such as economic relations; and fourth, power is *multidirectional*: Foucault repeatedly emphasizes that power does not flow exclusively from the ruling to the ruled but comes from below as well, and is thus different from domination.

Foucault thus develops a strategic understanding of power. It finds a first important expression in the notion of strategic games as ubiquitous form of power that flows in all directions: 'Between every point of a social body, between a man and a woman, between the members of a family, between a master and his pupil, between everyone who knows and everyone who does not, there exist relations of power [...] ' (PK: 187).[13]

Power is thus not identical with social, political or economic relations and neither determined by these, but crosses all other relations and has complex relations with these: 'Power relations are the immediate effects of the divisions, inequalities, and disequilibria which occur in the other types, and are the internal conditions of these differentiations; relations of power are not in superstructural positions, with merely a role of prohibition or affirmation; they have, where they come into play, a directly productive role' (Foucault 1998: 94).

Describing power as ubiquitous and multidirectional is not meant to claim equality between the members of society: it is in particular the mechanisms of exploitation and domination that lead to inequalities in power relations, and clearly define the above and below within these relations. However, a power relation always effects on both sides. Those who exercise power have to adapt their position as well, and power has thus effects on the constitution of subjectivity on both sides of a power relation (Dreyfus and Rabinow 1982).

Within strategic games, power is exercised not as domination or force, but acts on the behaviour of others by structuring their field of possible actions, and thus presupposes a minimum degree of freedom: '[W]ithout the possibility of recalcitrance, power would be equivalent to a physical determination' (Foucault 1982: 221). Power makes things more or less probable, 'it incites, it induces, it seduces, it makes easier or more difficult; [...] it is nevertheless always a way of acting upon an acting subject or acting subjects' (ibid. 220). Consent and violence, accordingly, can be the means, but not the principle or nature of power.

---

[13] This multidimensionality of power is the reason why Foucault refuses to a theory of power as well, or at least emphasizes that his analyses should not be confused with such a theory, because 'if power is in reality an open, more-or-less coordinated [...] cluster of relations, then the only problem is to provide oneself with a grid of analysis which makes possible an analytic of relations of power' (Foucault 1980a: 199). Instead of elaborating a theory of power, 'one needs to be nominalistic, no doubt: power is not an institution, and not a structure; [...] it is the name that one attributes to a complex strategic situation in a particular society' (Foucault 1998: 93).

Government as the conduct of conduct

Foucault suggests that the appropriate notion to capture the acting on the action of others is government: 'The relationship proper to power would [...] be sought not on the side of violence or of struggle, nor on that of voluntary contracts (all of what can, at best, only be the instruments of power) but, rather, in the area of that singular mode of action, neither warlike nor juridical, which is government' (EW3: 341). While strategic games describe the ubiquitous power relations between all members of society, government refers to the more organized ways of shaping the behaviour of others.

The understanding of power as government derives from the analysis of biopower and dispositifs of security. Obviously unhappy with the notion of biopower that has life as its centre, Foucault suggests that knowing the characteristics of a population has become relevant not only to address immediate life related phenomena, such as birth or mortality rates, but for governing other phenomena as well. In his 1978 lectures, he thus redefines biopower as *dispositifs* or *technologies of security.*

The hypothesis is that this new form of power neither acts in the juridical code of permission/prohibition, nor in the disciplinary code of good/bad, but rather as 'conditional imperative' or 'tactical pointers' (Foucault 2007: 3). While the operation of the law and disciplines is based in a distinction that is established in advance and external to the object of government, the dispositif of security develops its criteria for intervention from the object itself: 'it prescribes not by absolute demarcation between the permitted and the forbidden, but by the specification of an optimal within a tolerable bandwidth of variation' (Gordon 1991: 20).

In the case of delinquency, for instance, juridical and disciplinary mechanisms forbid certain actions or aim at shaping the individual's behaviour towards a distinct norm. The security dispositif, on the contrary, seeks to understand and regulate the phenomenon in its entirety, and in relation to different factors. Given that both crime itself and combating crime bears costs for society, it aims at an optimal rate of delinquency, and thus at the normalisation of the population around a criterion that is derived from its characteristics.

The security dispositif, however, does not replace the law and disciplinary mechanisms. Rather, it changes their form and correlation, 'taking up again and sometimes even multiplying juridical and disciplinary elements and redeploying them within its specific tactic' (Foucault 1998: 9).

Though not claiming a historical evolution and the substitution of juridical and disciplinary mechanisms with security mechanisms, Foucault seeks 'to investigate whether there really is a general economy of power which has the form

[of], or which is at any rate dominated by, the technology of security' (Foucault 1998: 11).

The security dispositif is not so much a technique of power, but a form of governmental rationality that manifests itself independently in different locales, at different times and in different fields of society. It shifts the focus of governmental intervention from the security of the sovereign, in form of the control over a territory, to the security of the population.

Security mechanisms address the population and thereby create it as an object of government. This distinct type of power in modern western society is captured in the notion *government*, as the 'right way of arranging (*disposer*) things in order to lead (*conduire*) them [...] to a "suitable end", an end suitable for each of the things to be governed' (Foucault 2007: 99).

These things can be natural environments or material goods as much as institutions, customs, or ways of thought; Foucault, however, focuses on the government of the population, and describes this as the conduct of conduct as well.

Conduct comprises both to lead others and to behave within a more or less open field of possibilities (ibid.). Instead of limiting the freedom of the individual, government works through and makes use of this freedom, and thereby constantly (re-) produces it. The population plays a role not only as the object of intervention, but in its subject qualities as well: government works through the desires and interests of the individual and the population as collective body. 'To govern, in that sense, is to structure the possible field of action of others' (Foucault 1982: 221).

Government and strategic games are both relational forms of power, where power is exercised in both directions with the objective to change the behaviour of the other(s). Nevertheless, they are distinct in important ways. Whereas strategic games describe the power relations between individuals, government is the coordinated way of acting on and through the freedom of the governed, it is 'the regulation of conduct by the more or less rational application of the appropriate technical means' (Hindess 1996: 106). Foucault locates the different forms of power within a continuum that has as its extreme points strategic games and states of domination, and places government between these ends.

2.2 The analytics of government and the role of the state

The notions of strategy and strategic games introduce a dynamic perspective on power, as the relations between individuals or groups can move in both directions on the power continuum. The objective of the adversaries in strategic games is to transform these games into a more stable situation that establishes a relation of subordination. 'A relationship of confrontation reaches its terms, its final moment (and the victory of one of the two adversaries) when stable mechanisms replace the free play of antagonistic relations. Through such mechanisms one can direct, in a fairly constant manner and with reasonable certainty, the conduct of others' (Foucault 1982: 225).

A situation of government thus is characterised, on the one hand, by a relation of subordination: it is clear in which direction power is primarily exercised (though, as has been emphasized before, the governing person is also affected by the exercise of power). But government must entail, on the other hand, a minimum degree of freedom on the side of the governed as well: Further extending the stable relationships of government can result either in a state of domination by constraining the freedom of the subordinated, or in a return of confrontation if the subordinated resist this limitation of freedom.[14]

2.2.1 The state as effect of governmental rationality

In this way, Foucault relates the different forms of power to each other: 'the idea of government as strategic codification of power relations provides a bridge between micro-diversity and macro-necessity' (Jessop 2007: 39). However, it remains unclear how the dispersed strategic games coordinate to more organised forms of government, and how social cohesion takes place. While Foucault 'is correct to problematize a descending concept of power, in which the concrete deployment and strategies of power are the manifestations of some global logic, the precise linking between the local and the global is not fully theorized' (Howarth 2000: 78).

Genealogy of the modern state

This is a direct consequence of not conceptualising the institutions of the state and the law as the origin of social organisation, and based ultimately in the rejection of a juridical conception of power. But while Foucault claims that 'in political thought and analysis, we still have not cut off the head of the king' (Foucault 1998: 88), he does not seem to provide an alternative model to explain social

---

[14] This dynamic of confrontation and power relations makes states of domination an important phenomenon for understanding the dynamics of societies, as 'they manifest in a massive and universalizing form, at the level of the whole social body, the locking of power relations with relations of strategy and the results proceeding from their interaction' (Foucault 1982: 226).

cohesion. The question that remains then is: 'How is it possible that his headless body often behaves as if it indeed had a head?' (Dean 1994: 156).

Honneth (1984) claims that Foucault's account of power shifts from a strategic theory of agency, in which power is constantly created and transformed within confrontations, to a systems theoretical account of society in which power is entrenched within institutions (in the case of disciplines) or state apparatuses (in the case of biopower), without making explicit this change. Like others, Honneth suggests to explain the cohesion of society and the institutionalisation of power relations through processes of communication, based on 'rational communicative action grounded on a universal reason' (Howarth 2000: 83).

However, Foucault has a conflictive understanding of societies and societal cohesion, which resembles Poulantzas' notion of state power as a material condensation of power relations or force relations in society. The moment of agreement and the acceptance of power relations it thus what requires explanation, instead of being the explanation for social cohesion.

Nevertheless, Honneth's critique points to an important problem in Foucault's work: He refuses to conceptualise the institutionalisation of (state) power, though he leaves no doubt that the state apparatuses play an important role as the end point of power relations (Foucault 2008). Power entails not only the various forces that inhabit a domain, but also 'the strategies in which they take effect, the general design or institutional crystallization of which takes shape in the state apparatus, in the formulation of the law, and in the social hegemonies' (Foucault 1998: 92-93).

Foucault addresses the question of state power not through a theory of the state, but within the perspective of historically distinct forms of government. He emphasizes that he is not interested in a definition or theory of the state, as the state has no essence and only exists as a historically concrete ensemble. It is thus possible, he claims, 'to place the modern state in the general technology of power that assured its mutations, development and functioning' (Foucault 1998: 120). The question, in other words, is 'how power relations historically could concentrate in the form of the state – without ever being reducible to it' (Lemke 2002: 59).

As in the analysis of institutions, Foucault's objective therefore is to address the question of the state from an exterior point of view, that is, to understand the formation of the state from the practices of government (Foucault 2008). He seeks to show how state apparatuses, domains of knowledge and governmental practices develop within a general tactic of power, an 'absolutely global project [...] which is directed toward society as a whole'.

The *governmentality* concept provides this perspective on the state from outside, by describing the interplay of governmental rationalities and forms of intervention as the background against which the state emerges in its concrete form. Governmentality is '[t]he ensemble formed by the institutions, procedures, analyses and reflections, the calculations and tactics that allow the exercise of this very specific albeit complex form of power, which has as its target population, as its principal form of knowledge political economy, and as its essential technical means apparatuses of security' (Foucault 1998: 108).

This is not to claim that the state was invented and brought about by modern forms of government, but that its current form has evolved within a particular approach to governing the population. Obviously, therefore, governmentality refers to particular societies at a certain time only. It results from the 'tendency which, over a long period and throughout the West, has steadily led towards the pre-eminence over all other forms (sovereignty, discipline, etc.) of this type of power which may be termed government, resulting, on the one hand, in formation of a whole series of specific governmental apparatuses, and, on the other, in the development of a whole complex of saviors' (Foucault 1998: 108).

Rationalities and technologies of government

Through this latter part of the definition, the status of the governmentality remains unclear: It is used either as an ahistorical concept or a historically specific category, both by Foucault and in the governmentality studies. To avoid difficulties with this unclear status, the analysis in this book builds on the analytical perspective that Foucault introduces with the governmentality concept, without making reference to the notion itself.

Governmentality describes the link between (governmental) knowledge and practices and is thus immediately relevant for understanding the power-knowledge nexus. The neologism 'governmentality' reflects the fundamental connection between acting or governing (*gouverner*) and thinking (*mentalité*).[15] As a particular combination of rationality and technologies, governmentality is used by Foucault almost interchangeably with *art of government* and 'indicates that it is not possible to study the technologies of power without an analysis of the political rationality underpinning them' (Lemke 2002: 50, Gordon 1991).

Rationality refers here to the consistency of rules, procedures and ways of thought, and the conditions that allow addressing a particular problem at a particular time. In that sense, rationality refers less to reason and more to historical practices in their context (Lemke 1997). The rationality of a governmental prac-

---

[15] Though Foucault apparently did not have this in mind but derived *governementalité* from the French adjective *gouvernemental*, to contrast his understanding of power from sovereignty (*souveraineté*) (Lemke 2007).

tice is the 'thinking about the nature of the practice of government [...] capable of making some form of that activity thinkable and practicable both to its practitioners and to those upon whom it is practised' (Gordon 1991: 3).

Governmental *technologies* are not neutral instruments, then, as they carry the rationality within which they have been developed, and thus permit extension of rationalities through time and space. Governmental rationalities and technologies are performative in creating, transforming and mobilising forms of political authority and subjectivity, and in creating the objects of government they address. The governmentality concept enables to analyse of governmental rationalities through governmental practices, as the latter make visible the former (Bulkeley, Watson et al. 2007).

The analysis of governmentality thus requires, on the one side, a description of its rationality and forms of intervention – or *art of government* – as the context in which the concrete form of state power evolves. In that sense, Foucault describes liberal governmentality as an explicit critique towards excessive regulation and state intervention. Liberalism rationalises the role of the state, it is 'in Kantian terms [...] a *critique of state reason*' (Gordon 1991: 15). Rather than a theory only, liberalism is 'an instrument for the criticism of reality' that defines the appropriate spaces and forms of intervention (Foucault 1981: 356).

The other side of analysing governmentalities is a genealogical perspective on the emergence of a particular governmental constellation. In the case of liberal governmentality, Foucault highlights the historical context and institutional settings that blocked the emergence of a liberal art of government. Though the ideas of an art of government that focuses on the population were already formulated in the 16th and 17th century, it 'remained imprisoned, as it were, within the forms of the administrative monarchy' (Foucault 2007: 152).

Just as a number of political, historical and institutional reasons blocked the development of liberal government, its release is also situated within a number of general processes, such as population growth, increases in agricultural production, and in particular 'the emergence of the problem of the population' (104). This is not meant to claim a causal relation between environmental changes and changes to governmental practices, but rather describes 'a quite subtle process, which we should try to reconstruct in detail, in which we can see how the science of government, the refocussing of the economy on something other than the family, and the problem of the population are all interconnected' (ibid.).

A governmentality perspective on the emergence of state formations thus combines archaeology and genealogy: it demonstrates the interplay of changes in the context of governmental practices, the formation of new domains of knowledge and objects or spheres of government, and institutional changes. The emergence of liberalism, in that sense, is located in a 'deep historical link' between a new art of government, the movement that brings about the population as a field of intervention, and finally the 'process that isolates the economy as a specific domain of reality, with political economy as both a science and a technique of intervention in this field of reality' (ibid. 108).

2.2.2 Emergence and change of governmental rationalities

Through the study of historically distinct governmentalities, Foucault claims to offer a 'genealogy of the modern state and its different apparatuses [...] on the basis of a history of governmental reason' (Foucault 2007: 354). However, the governmentality perspective does not account, in a systematic manner, for how certain constellations emerge and become dominant. The challenge, therefore, is to develop an analytical perspective that allows understanding the emergence and change of historically distinct rationalities, and the interplay with practices and institutions in that respect.

We can achieve this by introducing two concepts that gain less attention in the reception of Foucault's work. The notion of problematisation highlights the role of thought as an active intervention in the emergence of historically distinct rationalities and in responding to problems in the environment of governmental practices. The dispositif concept describes the emergence of a new governmental domain and rationality as the loose coupling of various strategies, without the need for a hegemonic force or a coordinating centre.

Problematisation by thought

In the beginning of this chapter we saw that Foucault introduces the notion of problematisation as the concept that characterises the analytical perspective of his studies. The objective of analysing problematisations is to 'define the conditions in which human beings "problematize" what they are, what they do, and the world in which they live' (Foucault 1986: 10). It thus describes not only the changes to governmental rationalities, but the process in which rationalities are formed: It emphasizes both thought as a driver of historical change and the structural conditions that enable and constrain processes of thought.

Foucault emphasizes that problematisations are linked to changes in social practices and relations and the emergence of phenomena such as madness, mental illness or delinquency (Foucault 2005b): 'I have tried to show that it was precisely some real existent in the world which was the target of social regulation at a given moment' (Foucault 2001: 171-2). A problematisation is 'an "answer" to a concrete situation which is real' (Foucault 1986: 10).

However, a problematisation is not the necessary effect or consequence of such a 'real situation'. Social, economic, or political processes play an important role in making something the object of thought, but do not determine the process of problematisation: 'They can exist and perform their action for a very long time, before there is effective problematisation by thought. And when thought intervenes, it doesn't assume a unique form that is the direct result or the necessary expression of these difficulties' (Foucault 2005b).

Problematisations, therefore, contain a moment of necessity and a moment of creativity. On the one hand, a problematisation is 'an answer given by definite individuals' and 'always a kind of creation' (ibid.): It is in particular Foucault's 1978-9 lectures that give new emphasis to thought as an active response to historically situated problems, and thus as a way of transforming power-knowledge constellations (Collier 2009).

On the other hand, these processes of thought are constrained by the structural conditions in which they take place: Problematisations are thought in historical context, and highlight the contingency of a particular governmental rationality this way.

A problematisation is not the concrete strategy to solve a problem, but 'develops the conditions in which possible responses can be given; it defines the elements that will constitute what the different solutions attempt to respond to' (Foucault 2005b). A problematisation transforms obstacles or difficulties into problems to which political solutions can be formulated. It defines the object of government and the limits of the political-epistemological field for the political practices that address this object: A given problematisation makes possible quite diverse responses in form of political strategies and practices and is the common ground on which these responses are formulated.

Analysing problematisations thus requires, on the one hand, a critical examination of how the different solutions to a problem have been generated, and on the other hand an examination of the common rationality in which they are based. We saw that in the political context, Foucault refers to rationality of government in a very similar way as *art of government* (Gordon 1991). Political or governmental rationality is the knowledge that informs the practice of government. Rationality determines what is seen as right and wrong or true and false in a certain epoch; it thus also highlights the coherence of the elements within a certain constellation (Suárez Müller 2004).

Strategies without a subject

The dispositif concept adds another perspective on the emergence and change of governmental rationalities, focusing on the activities and practices that contribute to the rationality or strategy of power. In the WK and his 1978 lectures, Foucault describes the emergence of the dispositifs of sexuality and security, and how they create their objects in the interplay of new discourses, institutions and practices.

The dispositif concept thus serves to describe the formation of a power-knowledge constellation in a particular historical context. The dispositif is

'a thoroughly heterogeneous ensemble of discourses, institutions, architectural forms, regulatory decisions, laws, administrative measures, scientific statements, philosophical, moral and philanthropic propositions–in short, the said as much as the unsaid. Such are the elements of the apparatus. The apparatus itself is the system of relations that can be established between these elements' (Foucault 1980a: 194).[16]

Similar to a problematisation, the starting point for the emergence or evolution of a dispositif is a change in the environment of governmental practices: the dispositif, in that sense, is 'a sort of [...] formation which has as its major function at a given historical moment that of responding to an *urgent need*. The apparatus thus has a dominant strategic function. This may have been, for example, the assimilation of a floating population found to be burdensome for an essentially mercantilist economy [...]' (Foucault 1980a: 195).

This strategic function defines the 'system of relations' or net that is tied between the elements of a dispositif; it relates discourses, institutions and practices to each other and directs them towards a common objective.

This strategy is distinct from both the intentional actions of individuals and the tactics of governmental technologies: individuals act intentionally and with a certain purpose in their personal realm and sometimes can reflect the local tactics of the technologies of government they deploy; however, they do not necessarily know, let alone understand, the effects that these practices have: 'People know what they do; they frequently know why they do what they do; but what they don't know is what they do does' (Foucault, cf. Dreyfus and Rabinow 1982: 187).[17]

---

[16] There is no proper translation for the French *dispositif*: while it is sometimes coined as 'apparatus', Dreyfus and Rabinow (1982) suggest 'grid of interpretation'; however, they acknowledge, this neglects the fact that for Foucault, the dispositif is not only an analytical tool but also the relation between the elements of the dispositif.

[17] Foucault substantiates this point drawing on his empirical studies on the prison: while the guards in their daily practices are led by both the institutions of the prison and a variety of personal motivations

It is not possible to identify an actor or group of actors as the origin of a strategy of power, it does not require a hegemonic force or a coordinating centre (Jessop 2005). Rather, the strategy develops in the loose coupling of a variety of practices and programmes that do not necessarily pursue the same – intentional – objective. This is what is behind Foucault's claim that the strategies of power are at once intentional and non-subjective, that 'it is often the case that no one is there to have invented them, and few who can be said to have formulated them: an implicit characteristic of the great anonymous, almost unspoken strategies which coordinate the loquacious tactics whose "inventors" or decision makers are often without hypocrisy' (Foucault 1998: 95).

We can understand this better by means of an example. Foucault describes the interplay of changes to the rent and wage payment systems and the introduction of housing and educational programmes in France in the 19th century, which made it more difficult for industry workers to move from one factory or town to another. These practices added to another in a way 'that results in a global, coherent strategy, of which it would be impossible to say, however, who developed it' (Foucault 1978b: 133, own translation).

In a certain sense, the strategic function of a dispositif is comparable to the role the episteme: as the rule of formation not for scientific knowledge, but for political practice.[18] Suarez Müller (2004) describes the episteme as *theoretical rationality* and the dispositif as *strategic rationality* of an epoch, or as the rule governed rationality of a (power) constellation (see also Deleuze 1992). The rationality of an epoch is strategic to the degree that it enables, or makes possible, discursive, normative and institutional practices, that, vice versa, constitute and maintain this rationality. The dispositif can thus be understood as the consistent strategy that forms from the institutional, discursive and normative integration of a society.

It is crucial to see that the emergence and evolution of a strategy does not leave the elements of the dispositif untouched. The dispositif re-organises objects, subjects and modes of government towards the strategic objective. This entails a certain manipulation or reorganisation of the forces within a given constellation, 'a rational and well-tuned intervention into these power relations, to

---

and more or less understand the objectives of the disciplinary mechanisms in the prison, neither they nor the ones who put in place these mechanisms have intended and understand the global strategy of the disciplinary regime and its effects. These can only be understood when looking at the role of disciplinary mechanisms within whole society.

[18] Foucault suggests that a dispositif is the more general case of an episteme as it contains discursive and non-discursive elements, or vice versa, that the episteme is the specific discursive form of a dispositif. On the distinction between discursive and non-discursive elements, see the section on ontology below.

extend them into a certain direction, to block or stabilize them, or to make use of them for a certain purpose' (Foucault 1978b: 122/123, own translation).

Jessop (2005) suggests that this manipulation resembles a process of strategic selectivity: the strategy limits the flexibility of the different elements and micro relations of power by integrating them into larger systems.

The relation between the dispositif and its elements is not one-sided, however: Changes to the elements induce changes to the dispositif as a whole as well. Foucault describes this process in two ways.

First, through the strategy as the net or connection between the elements of a dispositif, these elements are mutually enabling, constraining and reinforcing. Due to this *overdetermination*, every irritation or change to one of the elements requires the adjustment of the others, and ultimately to the dispositif itself (Foucault 1978c).

Second, the dispositif changes in a perpetual process of *strategic elaboration or completion*: The strategy does not determine the practices within a dispositif, so that some practices are in tension with the dispositif or not yet addressed. The strategy therefore evolves whenever new elements and relations of forces become part of the dispositif. It is in that sense, for example, that the strategy of the penal systems was modified and new techniques were developed to address the milieu of delinquency that had emerged out of the penal system

2.2.3 Analysing discourse from (governmental) practices

We can understand the dispositif concept as a way of concretising a genealogical research strategy: Rather than to approach a political field from the institutions or apparatuses for the regulation of a certain object, the dispositif perspective suggests to start from the practices that are aim at a similar object and objective. As a *grid of interpretation* that is constructed within the analysis, the dispositif enables the genealogist to account for the emergence of practices and institutions within a broader and critical perspective (Howarth 2000).

Foucaultian genealogy, in that sense, aims at an interpretative analytics of the practices that have become seen as 'altogether natural, self-evident and indispensable' (Foucault 1991: 75). Shaking this self-evidence and necessity is not meant to describe a particular set of practices as arbitrary; rather, it is meant to reveal how they are bound to a certain historical context of emergence, and thus embody a rationality that is not reducible to the intentions and understanding of the acting individuals or institutions. This rationality 'has no essence, no fixity, no hidden underlying unity. But it nonetheless has its own specific coherence' (Dreyfus and Rabinow 1982: 124).

Such an approach is interpretative but not hermeneutic as it neither remains on the level of everyday understanding nor claims to reveal a hidden meaning behind the practices; rather it is 'a pragmatically guided reading of the coherence of the practices of the society' (ibid. 125).

Combining regimes of practices and a dispositif perspective

Dean (2010) takes the intelligibility of practices as the starting point for developing an *analytics of government*. It focuses on how practices of government make visible objects of government, and how this is related to processes of knowledge and identity formation. Dean´s approach is helpful for the analytical purposes of this book as it combines and systemises a governmentality perspective with insights from the problematisation and dispositif concept.

Following the understanding of government as the 'conduct of conduct' (Foucault 1982: 220-1), the analytics of government focuses on modern forms of government or governmentality. Government here is 'any more or less calculated and rational activity, undertaken by a multiplicity of authorities and agencies, employing a variety of techniques and forms of knowledge, that seeks to shape conduct by working through the desire, aspirations, interests and beliefs of various actors, for definite but shifting ends and with a diverse set of relatively unpredictable consequences, effects and outcomes' (Dean 2010: 18).

The analytics of government thus follows an objective that is very close to what Foucault addressed with the notion of problematisation: to understand and interpret how men govern themselves through the production of truth. For Dean, the objective is 'to make intelligible the limits and potentials of who we are and have become, and the ways our understanding of ourselves is linked to the ways in which we are governed [...] under forms of knowledge postulated as truth by various authorities' (ibid. 14).

Dean defines the object of inquiry as regimes of practices that, close to the definition of government, are 'the more or less organized ways, at any given time and place, we think about, reform and practise such things as caring, administering, counselling, curing, punishing, educating and so on' (ibid. 31). Constituted and delimited by the practices that aim at a particular object of government, regimes of practices are, like the dispositif, composed of heterogeneous elements that have diverse historical trajectories; they are 'historically constituted assemblages' (ibid. 40).

A regime of practices is thus not identical with a certain institution or policy field, but can cross these borders and take effect within several institutions or fields. This non-identity is important for analytical purposes: institutions are usually characterised by a certain set of principles or objectives. Practices or regimes of practices, to the contrary, 'possess up to a point their own specific regularities, logic, strategy, self-evidence and reason' (Foucault 1991: 75) that are not reducible to intentional formulations.

The dispositif concept helps to specify this relation: it is the strategic function of the dispositif that links its elements, and thus defines the boundaries of a dispositif; whether something is an element of a certain dispositif depends on its contribution to the strategy. The dispositif or regime of practices perspective is thus particularly helpful for analysing the power-knowledge relations in domains of knowledge and acting that are not dominated by a single institution or formation of knowledge (Traue 2010).

The logic of a regime of practices is close to political rationality as contained in the notion of governmentality, in the sense of the practical knowledge that informs the practices of government. It includes the different ways 'in which these institutional practices can be thought, made into objects of knowledge, and made subject to problematizations' (Dean 2010: 31). Likewise, the knowledge entailed in the dispositif has been described as programmatic knowledge or the operating manual for the elements contained in it.

The logic of a regime of practices is thus the particular way of thinking about a problem, often drawing on a specific theoretical or technical knowledge, but does not imply hegemony of this way of thought.[19] As a practical form of knowledge, the logic or rationality is embodied in language or techniques. The governmentality perspective highlighted that technologies of government carry with them a particular rationality. The analysis of government rationality is thus an analysis of thought made practical and technical, a materialist analysis in the sense of 'materialism [...] concerned with thought' (ibid. 41).

As the analysis grants to practices a reality and logic of their own, it must thoroughly describe these practices, to identify the rationality as the common ground that makes these practices possible and directs them towards a common objective. It is crucial to distinguish between the intentional programmes of individual actors that seek to invest regimes of practices with particular purposes, and the logic of a particular regime of practices that cannot be read off these programmes, but only from the regime of practices itself: 'The strategic logic of a regime of practices can only be constructed through understanding its operation as an intentional but non-subjective assemblage of all its elements' (Dean 2010: 32).

---

[19] Dean points out that his understanding of government rationality or mentality, carrying the idea of thinking as a collective activity, is similar to an understanding of sociologists like Durkheim of a collective mentality as 'a collective relatively bounded unity [...] not readily examined by those who inhabit it' (Dean, 25).

This requires including in the analysis all the objects, practices and processes that contribute to a particular problem understanding and thereby inform, support, and shape a particular set of practices.[20] These elements are not analysed as such, however, but in the way they are rationalised as part of governmental practice and are meant to contribute to a certain objective. This highlights that objects, subjects, and the places, forms and techniques of government are formed and transformed within the processes of rationalising and exercising government, as are power relations.

The analysis of practices thus combines archaeology and genealogy in describing how diverse elements and processes are bundled together and assembled into more stable forms of institutional or governmental practice (Dean 2010); it thus also operationalises the problematisation perspective, demonstrating how these processes constitute a domain of objects and practices and define the rules for governing these objects (Foucault 2005b).

A heuristic for analysing governmental thought and practice

The heuristic that Dean develops for the analytics of government builds on how Foucault and Deleuze have used and systemised the dispositif concept. Describing the emergence of dispositifs of security, Foucault highlights different ways in which the population is constituted as an object of government, including: its visibility, through statistics that map the characteristics of a population; its appearance in domains of knowledge such as political economy, which open new fields of intervention; and new or changing forms of acting on this object, such as the technologies of security that address the wellbeing of the population (Foucault 1998).

Deleuze has systemised the analytical potential of the dispositif concept by describing it as a 'multilinear ensemble' (Deleuze 1992: 159) that produces its objects along these lines or vectors. These lines do not describe or outline homogeneous systems with fixed objects, subjects, and concepts, 'but follow directions, trace balances which are always off balances, now drawing together and then distancing themselves from each other' (ibid.). Knowledge, power and subjectivity within a dispositif are thus never fixed and clearly contoured; they are always in flux and effect on each other (Ewald 1978).

Deleuze describes the lines along which objects are shaped as (1) visibility: each apparatus has its own way of structuring light, and therefore of distributing

---

[20] This can include such diverse elements as administrative structures, the (re-)organisation of government departments, expert training and education, information systems, grading and certificate systems, and not least the various forms of interaction between the governing and the governed.

the visible and the invisible, making some objects appear and others disappear; (2) affirmations/enoncés as the curves that define the status of these objects, and are inherent in sciences, art genres, or the law; (3) lines of forces, which 'rectify' the preceding curves and fill in the space between the different lines, 'acting as go-betweens between seeing and saying and vice versa' (Deleuze 1992: 160, Ewald 1978) and thus link things with knowledge and language; (4) and finally, lines or processes of subjectification that both affirm and transcend the other lines and prevent social apparatuses from becoming 'locked into unbreakable lines of forces which would impose definite contours' (ibid.).

Dean adapts the description of lines or vectors within dispositifs or social apparatuses as four dimensions for the analytics of government.

- Regimes of practices first are characterised through and produce a certain field of *visibility*. This includes maps, charts and other media which make the object of government visible and understandable, indicate what problems are to be solved, show the connections between different agents and locals, and mark relations of authority and obedience. It is important to see that different regimes of practices can make different aspects of the same object visible.
- Second, regimes are characterised by a specific *technical dimension of government*, that is, by the tactics, techniques, or procedures they deploy. Dean (2010) emphasizes that the notion of technology of government refers to organised forms of government practice only. This highlights the need to distinguish between technologies and techniques of government. Technique refers to mere technical instruments whose effect depends on the context of their deployment. Foucault, in that sense, introduces the panopticon as a technique that can be used in various contexts and for various purposes, while being ambivalent regarding the objective of its use. 'It is only when it "invests" and undermines other institutions that it takes on its own momentum' (Dreyfus and Rabinow 1982: 192). Technologies, to the contrary, link several techniques and direct them towards a strategic objective (Traue 2010: 240).
- The third characteristic of a regime of practices is its *episteme of government* (Dean 1995), that is, the forms of knowledge, thought and rationality that arise from and inform the practices of governing. The welfare state concept and neo-liberalism are such a form of rationality rather than a definite set of practices.
- Fourth and final are the *forms of identities* through which governing operates, on both the side of the governing (as experts, politicians, professionals) and the side of the governed (workers, consumers, pupils). Identities are

60

formed within practices and programmes; we can ask what forms of conducts are expected from the respective identities, what rights and duties they have, and how these duties are enforced and rights ensured. These processes of identity formation do not form new subjects, but ascribe certain capacities, statuses or qualities to individuals.

The last part of the theory chapter refines this heuristic for analysing the emerging investment dispositif in climate governance, in particular with respect to the archaeological and genealogical dimension of research.

2.3 Operationalisation

The notions of problematisation, dispositif and regime of practices provide us with the theoretical concepts to operationalise a governmentality perspective, and to highlight the mutual constitution of practices and rationalities of government. The first part of this section addresses the ontological and epistemological implications of this research strategy, and builds on this in developing the methodology of this book. The second part gives an overview of the research and refines the theoretical concepts for the different steps of analysing the Stern Review and the climate investment discourse.

2.3.1 Ontology, epistemology, methodology

The use of concepts like discourse, logic and rationality, and the definition of a methodology and research strategy building on these concepts, raises important ontological questions. One of the strengths of Howarth and Glynos' (2007) contribution is to make explicit the ontological underpinnings of a research strategy that aims at analysing the discursive production of meaning, and to relate the ontological, epistemological and methodological questions of such an endeavour. The following builds on their discussion of a poststructuralist methodology in form of retroduction and articulation that aims at a middle way between positivism/causal explanation and hermeneutics, and uses the model of a single explanatory chain to define the research steps of this project.
Ontology: discourse, contingency

Describing the ontological assumptions and presuppositions of a research project is essential as every empirical and normative investigation mobilises underlying presuppositions that have important consequences for epistemology and methodology, and therefore, ultimately, for the results of a research project. Ontology, in that sense, addresses the basic theoretical assumptions that mark important differences, for example, between positivism and post-positivism. Speaking with Heidegger, ontological enquiry concerns the categorical preconditions for the objects and for their investigation, that is, is questions the basic concepts, while ontical enquiry focuses on the specific types of objects and entities within a particular domain (Howarth and Glynos 2007: 109).

The two central ontological assumptions of this study that are at the heart of poststructuralist theorising are the radical contingency of the social and the discursive production of meaning.

To specify what is meant by the discursive production of meaning, we can follow Laclau and Mouffe in describing all practices as discursive in the sense of articulation: 'We will call articulation any practice establishing a relation among elements such that their identity is modified as a result of the articulatory practice. The structured totality resulting from this articulatory practice, we will call discourse' (1985: 105).

This is not meant to deny the non-discursive existence of objects and a material world; however, all understanding and perception of objects, and in consequence their relevance within the social world, is always constituted in discourse. 'Ontologically things may appear to exist independently of thought or consciousness, but epistemologically they enter discourses as things we give names to' (Caldwell 2007b: 782).

There seems to be a fundamental difference between Laclau and Mouffe, who describe all practices as discursive practices, and Foucault, who repeatedly distinguishes between discursive and non-discursive practices. In separating the episteme as the central unit of archaeology from the dispositif as analytical grid in a genealogical research strategy, Foucault also refers to discourse as one element within the dispositif among others.

However, his remarks are not meant to question the primacy of discourse in the sense of the primacy of meaning to practices: what Foucault aims at with distinguishing discursive and non-discursive elements within the dispositif is rather to emphasize that it is not only text or spoken language that contributes to the strategy and thus the rationality of an epoch, but elements that are not articulated this way as well. What he rather refers to, then, is the difference between textual and non-textual practices. It is not discursive and non-discursive elements that are contained in the dispositif, but 'the said as much as the unsaid' (Foucault 1980a: 194).

This points to the inconsistent and changing way in which Foucault uses the notion of discourse: in its widest sense, discourse – and likewise knowledge – is synonymous with the rationality of an epoch, and thus contains linguistic as well as non-linguistic elements (Suárez Müller 2004). But sometimes Foucault uses discourse in a more narrow way, in the sense of particular domains of knowledge or scientific disciplines; discourse in this sense is synonymous with episteme. His use of the dispositif concept is thus not at odds with the ontological primacy of discourse but rather supports it. Foucault emphasises that the dispositif is defined through the elements that contribute to the rationality of an epoch, whether or not these elements play a role in public or scientific discourses of that time (Foucault 1978b). Against this background it makes sense to substitute the notion of discourse in the definition of the dispositif through *episteme* or *formation of knowledge* (Denninger, van Dyk et al. 2010); Laclau and Mouffe´s understanding of discourse, then, contains all the elements of the dispositif.

The second fundamental ontological assumption that we can find in poststructuralist theorising is (radical) contingency, defined by Laclau as the necessary openness of social structures. The notion of contingency ranges prominent in the Discourse Theory of Laclau or Laclau and Mouffe and those who deploy their approach (see, for instance, the contributions in Butler, Laclau et al. 2000). It is particularly important for the notion of radical democracy. Laclau claims that the only democratic society is one 'which permanently shows the contingency of its own foundations' (Laclau 2000: 86), that is, a society that is conscious of the non-necessity of its social configuration and the particular decisions that are taken.

The assumption of radical contingency, however, is held not only in poststructuralist theorising. In philosophy and sociology in general it highlights the general possibility of things being different. In that sense, the notion of contingency has an important role in the Systems Theory of Niklas Luhmann, for instance: 'Something is contingent insofar as it is neither necessary nor impossible; it is just what it is (or what it was or will be), though it could also be otherwise. The concept thus describes something given (something experienced, expected, remembered, fantasized) in the lights of its possibly being otherwise; it describes objects within the horizon of possible variations' (Luhmann 1995: 106).

For Howarth and Glynos, the notion of contingency is a necessary part of clarifying the ontological assumptions and commitments of social science research. this includes to not only specify what kind of objects exist in the world, 'but *that* they exist and *how* they exist' (Howarth and Glynos 2007: 11). The notion of contingency, in that sense, makes clear that objects and subjects are marked by an essential instability, and highlights the '*constructed* and *political* character of social objectivity' (ibid.).

To clarify this point, we can distinguish between *radical* or *ontological* contingency that characterises poststructuralist theorising, and *empirical* or *ontical* contingency that refers to the (accidental) instability of an object's identity: In the latter understanding, the contingency of objects is understood as possibly temporary, a contingency that may be overcome ultimately by a higher order process that fixes the meaning of an object. Radical or ontological contingency, to the contrary, assumes the impossibility of fixation, very much in the sense of 'structural undecidability' (Derrida).

The notion of contingency marks the *post* in poststructuralist theorising: unlike many other versions of structuralism, it rejects the assumption that social reality and social development are determined by structures. While it is still the structures that are at the centre of the analysis, these structures are understood as necessarily incomplete, and thus open to change: contingency refers to the fact that things can be and could have been different, that there is no necessity for the realisation of a particular social relation. It highlights as well that social reality is not shaped by one specific set of structures, such as economic structures in much of Marxist theorising, but result from different sets of structures (Diefenbach 1999).

From a discourse theoretical perspective, it is the discursive construction of objects and reality that opens a space for different understandings and interpretations. This is, however, not to be misunderstood as a free floating of meanings, as *everything is possible in discourse* – meanings are still closely related to structures, but nevertheless not determined by them. Contingency therefore makes room for agency not only in the constitution of – always historical – structures, an assumption that poststructuralists share with structuralists, but also in the sense of space for agency *within* these structures.

Through the notion of problematisation, Foucault highlights the role of thought in the way that historical conditions and structures take effect on human agency, and in particular on governmental practices. Thought describes the space between the 'real existent in the world which was the target of social regulation at a given moment' (Foucault 1985b: 115), and problematisation as 'an answer given by definite individuals' (ibid.). From contingency follows that there is always the possibility of alternative problematisations; the task of research therefore is to identify and describe the conditions that gave rise to a particular problematisation instead of others.

Epistemology: Beyond hermeneutics and positivism

The ontological assumptions of discursively produced meaning and radical contingency inform a research strategy that rejects both hermeneutics and positivism; it aims at a middle way between the relativism or particularism of actors' self-interpretations and the universalism and objectivism of positivist accounts. The objective is a form of explanation that respects the self-interpretation of social actors and is nevertheless not reduced to subjectivism, but allows for a certain degree of generalisation and thus critique.

The rejection of positivist universalism follows directly from the assumption of discursively produced meaning and radical contingency: as all identities are always precarious and the status of objects open to different interpretations, a direct causal relation between social facts is impossible. This implies a rejection of deductive forms of explanation; both in form of justifying or testing theoretical hypotheses through empirical data, and as a deductive nomological model that deduces an explanandum from general premises or a universal law.[21]

A possible alternative to deduction or induction is retroductive explanation. The retroductive moment is based in the assumption that the researcher cannot simply find, but has to actively construct a model for the causal mechanisms that account for the phenomena in question. Retroduction, therefore, breaks with the separation of a context of discovery and a context of explanation, that is, the positivist assumption that a theory can account for social phenomena independent from the context as long as it can be verified through testing.

Rejecting universalism this way is not meant to reject the explanatory power of theories in general; rather, retroduction seeks to describe the way plausible hypotheses are generated, and thus constantly moves between theoretical assumptions, hypotheses and empirical data, and adjusts the hypotheses within the process of research.

Howarth and Glynos (2007) embrace retroduction as a step towards post-positivist social science practice and an appropriate *form* of explanation, but they reject causal mechanisms as its *content*: social science approaches like critical realism introduce causal mechanisms as a – weaker – alternative to causal laws. However, Howarth and Glynos dismiss this proposal, due to a residual positivism and the proximity to causal laws.

As an alternative content for retroductive explanation, Howarth and Glynos consider hermeneutical accounts, somewhat the other extreme of social science explanation. Drawing mainly on the work of Peter Winch and Charles Taylor, they describe hermeneutics as the 'practice of interpreting the self-interpretations of social actors in particular historical contexts' (Howarth and Glynos 2007: 50ff.). Acknowledging the convincing critique of positivist universalism, they nevertheless criticise that hermeneutics replaces it by different forms of particularism.

---

[21] Such a distinction seems to be blurred anyway: 'Explaining x is predicting x after it has actually happened. [...] Predicting x is explaining it before it has actually happened' (Hanson 1972: 41); this perspective is widely shared by positivists like Popper as well (Howarth and Glynos 2007).

The weakness of hermeneutics, in that sense, is to remain on the level of self-interpretations, and therefore to offer very limited explanatory power. Howarth and Glynos (2007) problematise in particular the impossibility to evaluate the interpretations of actors against any form of external position or criteria; this results either in a form of relativism that does not allow for any form of judgement, or, starting from a strong normative position, in normativism or normative particularism.

Rejecting both the causal law paradigm and pure hermeneutics, Howarth and Glynos (2007) rely on poststructuralist explanation, that is, they specify the post-positivist to a post-structuralist approach that conceptualises the transformation of social structures, and clarifies the relationship between social structures, political agency and power.

Poststructuralist reasoning integrates insights from both causal explanation and hermeneutics, while rejecting and replacing their problematic features: The motivations of actors to comply with a certain regime of practices or to fulfil a certain action can only be identified these actors themselves, but they do not necessarily reflect the social logics that induce these motivations. The analysis and interpretation must therefore take into consideration the structural contexts and conditions that can explain these actions, describing their institutionalisation, sedimentation, or transformation. The analysis of a set of practices must therefore include theoretical or generalised forms of explanation as well.

Methodology: retroduction and single explanatory chain

To define a research methodology that is consistent with the ontological assumptions of radical contingency and discursively produced meaning, and that chooses a middle way between hermeneutics and positivism, Howarth and Glynos build on retroduction as a *form* of explanation. Retroduction is an 'active process of problematization [that] involves the constitution of a problem – or an explanandum to use more traditional terms – which invariably results in the transformation of our initial perception and understandings' (Howarth and Glynos 2007: 34).

The starting point and first step for this form of an explanation, consequently, is a phenomenon that requires being rendered intelligible. Retroduction, in that sense, is a *problem driven approach* that acknowledges the active role of the researcher in constructing an explanation, as the puzzle is put together as a first step of the research. The second step is then to find an explanation for the phenomenon in question, in the specific form of retroduction as a constant movement between explanans and the potential explanandum. As a consequence of rejecting generalised explanations, the criteria for the justification of the explanandum are internal to the process of explanation as well, and ultimately consist in rendering the phenomenon intelligible convincingly.

Rather than subsuming empirical phenomena under a general theory or rule, retroductive explanation therefore takes the form of *articulation*. Articulation refers here to a form of identifying a particular empirical instance as theoretically significant by describing objects or practices as features of a particular logic. This inevitably contains a moment of naming and signification and, at least temporarily, fixes meaning. Articulation involves the creation of something new in that sense and is understood as redescription as well.

As an almost circular process between the hypothesis, empirical data and explanatory concepts, retroductive explanation contains a moment of theory construction as well, as the sense and meaning of explanatory categories are constantly transformed in the very process of their application. Taken together, the steps of explanation build a *retroductive circle* of *problematisation, interpretation* and *ontological projection*.

To specify this form of explanation towards a methodology for critical social science explanation, Howarth and Glynos (2007) suggest constructing a *single explanatory chain* that consists of the problematisation of empirical phenomena; the retroductive explanation of these phenomena; their interpretation in the light of existing theoretical and explanatory concepts; and critique in form of 'deconstructive genealogy' that discloses possible alternatives that have not been realised (ibid. 155).

Such an explanatory chain is an active, non-necessary drawing together of elements for the purpose of explanation, including both empirical data and theoretical concepts. In this process it is fundamental to ensure the compatibility of explanatory concepts. The first two steps of the explanatory chain of this study will be filled through the concepts of problematisation and regimes of practices/dispositif respectively, and described in more detail below; some further remarks are necessary here on the steps of interpretation and critique.

The interpretation or analysis of the practices in question requires drawing on general explanatory concepts, as self-interpretations alone cannot fully explain the motivations of actors. However, the explanatory concepts have to be refined according to the empirical context and, consequently, will not remain intact in the research process. This *reactivation* of theoretical concepts is meant to avoid subsumption under existing concepts. This does not prevent, however, any form of generalisation: using the results of the analysis, generalisations are possible against the background of a theoretical framework, for instance, by comparing different types of – already articulated – discourses according to the narratives that are contained within them.

The other critical issue is the question if and how an analytics based on poststructuralist theorising can provide a basis for *critique*. Combining Foucault and Derrida, Howarth and Glynos (2007: 155) suggest a critique in the form of 'deconstructive genealogy' that highlights the contingency of a social configuration by disclosing possible alternative framings that have not been realised and whose exclusion is necessary for the existence of a particular social configuration. This form of critique inevitably contains a normative claim, as problematising the constitution of a certain social configuration implies that it is worth contestation – this highlights the 'ineliminable connection between analysis and critique [...]. The very naming of a social logic already involves critical judgement' (ibid. 194).

2.3.2 Research process and methods

Drawing on the theoretical vocabulary and the poststructuralist methodology developed above, I can now describe in more detail the different steps of analysis in this book. One immediate consequence of retroductive methodology as an 'active process of problematization' (though a feature that is by no means exclusive to retroductive reasoning) is that the process of analysis is distinct from the presentation of results: The problem that is to be solved – or the phenomenon to be rendered intelligible – is constructed and specified within the process of analysis, and the elements that are required for its explanation can only be identified stepwise, too.

In the case of analysing a dispositif or regime of practices, for instance, the contexts that are relevant for explaining its emergence and significance can only be determined after a preliminary analysis of its practices. In presenting the results of the research, however, it makes sense to describe these contexts first, to clarify the general environment in which the practices under scrutiny are enacted.

A single explanatory chain for changes to the rationality of climate governance

Altogether, the research process went back and forth between theory, hypotheses, and the object of analysis. The following describes the single steps of the analysis in form of a single explanatory chain (SEC) that accounts for the phenomena at question. It shows how the Stern Review and the climate investment discourse are related, and reflects their significance for the greater context of climate change governance. It entails the steps problematisation, retroductive explanation, interpretation and critique. After a quick overview of these steps, I outline the analysis of the climate investment dispositif as the main empirical contribution of this book in more detail.

Step 1: Problematisation. Retroductive reasoning starts with the constitution of the research problem. In the case of this book, it is two phenomena or developments that require explanation: the importance of the Stern Review in climate change debates and the emergence of the climate investment discourse. It is the central hypothesis of this book that their emergence and significance are related to each other. The Stern Review on The Economics of Climate Change gained huge importance in public and expert debates on climate change and is often used to justify calls for enhanced climate protection; the question is what caused this massive importance, given that, as the authors suggest, the study does not offer radical new insights into the climate problem. The new emphasis on investment for climate protection seems be related to the perspective introduced by Stern. The climate investment discourse problematises established ways of governing the climate, as it suggests focusing on the creation of low-carbon economies rather than on emission reductions.

Step 2: Retroductive Explanation. The retroductive explanation addresses the two phenomena in question separately. The first part seeks to explain for the *significance of the Stern Review*, or the discussion around it, in climate change governance in recent years. The analysis treats the Review as an economic *problematisation* of climate change: Instead of prescribing a single strategy for addressing the climate challenge, it outlines the conditions under which it is possible to align climate change and economic objectives.

The analysis is thus not limited to the Review itself, to its insights and flaws, but combines archaeology and genealogy in analysing the rationality contained in the Review, and the contexts that can explain for its importance. Part of this analysis is on the one hand the secondary literature on the Stern Review: it highlights an important gap between the the tremendous importance the study gained in climate policy debates, and the reception within expert circles that see the study largely as one among many economic calculations of climate change. On the other hand, the analysis addresses the discursive environment that can explain for the significance of the Stern perspective on climate change, inter alia through interviews with two co-authors of the Stern Review. (The second part of retroductive explanation, concerning the climate investment dispositif, is the major empirical contribution of this book and is outlined in more detail above.)

Step 3: Interpretation and Critique. The interpretation of the empirical results as third step of the SEC brings together the economic problematisation of climate change and the investment turn in climate change governance. It starts from the greater role that is ascribed in both contexts to the state in managing the climate crisis and making markets work for climate protection, and asks whether and how this challenges the Kyoto markets and flexibility approach. It describes the government of investment as aligning climate protection with current market constellations, and the interests of the actors within these markets.

Building on this understanding, the last step of the SEC also formulates a critique of the climate investment discourse. It argues that the crucial question is not whether the investment turn means more or less state intervention, but what is made visible and invisible in the processes that ascribe new responsibility to the state in climate protection. Following the idea of *deconstructive genealogy*, it highlights alternative framings or understandings of the investment challenge in climate governance, focusing on: the way that climate protection is framed as a financial challenge; how the climate investment discourse does (not) address the role of the current financial markets for a low-carbon transformation; and how the finance turn frames climate protection as an issue of competition rather than cooperation.

Emergence and significance of the climate investment dispositif

The second part of retroductive explanation (and the major empirical contribution of this book) is the analysis of the government activities that address investment for climate protection. It understands this as the emergence of an investment dispositif in climate governance, and assigns the dimensions of Dean's heuristic for the analytics of government to two research steps that can be understood as more genealogical and more archaeological perspective on the government of investment.

The first, genealogical part of the research explains the emergence of investment for climate protection as an object of government, identifying the contexts in which this new object is addressed and the (discursive and non-discursive) practices that constitute it. A first analysis of climate change governance and discourse in recent years showed that it is in particular three contexts which are relevant here: the climate finance field, low-carbon strategies, and renewable energies policies.

The analysis therefore focuses on how investment for climate protection is constituted as an object of government within these contexts: It describes how the object is *made visible* in studies that calculate the need for low-carbon investment or display and make comparable the costs of different activities related to climate protection; and describes the *formation of knowledge* on investment as an object of government, and how this is related to the rise of new *subject positions* in climate governance.

Howarth and Glynos (2007) highlight that the contexts of emergence of new practices entail the conditions for their acceptability as well. To understand the formation, persistence and importance of the practices directed at investment for climate protection therefore requires identifying how different discourses or narratives give rise to and frame the investment challenge. This entails to highlight the differences between them but also how they interact, build on each other and converge around increasingly common understandings.[22] In this sense, the analytics of government takes a nominalist approach in showing how the concepts that are created in rationalising government activities enable these practices of government.

The second (and more archaeological) part of the analysis of the investment dispositif in climate change governance focuses on how different agencies aim at enabling and directing investment for climate protection, and thus at the *technical side* of the climate investment dispositif. This part of the analysis highlights that it is not only new climate finance and investment instruments that address investment as a new object in climate governance; rather, enhancing investment for climate protection has become a concern in different fields of climate governance, and plays a crucial role in the governance of carbon markets and the creation of a REDD mechanism for forest protection as well.

It is helpful to specify the technical side of government through the distinction between governmental techniques and technologies that has been made in the analysis of dispositifs: *Technique* refers to the technical dimension of practices such as techniques of calculation; *technologies*, to the contrary, are the organised and purposeful forms of governmental activity that link techniques and direct them towards a strategic objective (Traue 2010).

In that sense, we can distinguish between new technologies of government like Public Finance Mechanisms that direct techniques from other issue areas to the objective of raising investment for climate protection; and established technologies in climate change governance like the Clean Development Mechanism, which are adapted to the objective of raising climate related investment through new (technical) forms of government intervention.

The primary objective here is to identify and define the common rationality or strategy embodied within the practices directed at investment for climate protection. It is important therefore to emphasise the difference between the intentional *programmes* of government that seek to invest practices with particular

---

[22] With Foucault, discourse is understood here as particular formation of knowledge or episteme. The ontological difficulties related to Foucault's notion of discourse were addressed in the section on ontology (2.4.1).

purposes, and the *rationality* of government as the common ground of these programmes. This rationality or strategic logic is not fully reflected in every practice necessarily and 'can only be constructed through understanding its operation as an intentional but non-subjective assemblage of all its elements' (Dean 2010: 32).

This means, vice versa, that the form of the mechanisms and instruments directed at investment for climate protection is not determined by this common strategy or rationality neither; rather, the concrete form of these practices depend on the environments in which they emerge and are deployed, and thus also on the interests of the actors who deploy them.

The dispositif concept is thus well suited to describe the relation between different practices directed at investment for climate protection: It understands this as a loose coupling of strategies around increasingly common understandings and objectives rather than a moment of hegemony. A dispositif perspective, in that sense, highlights how various actors can relate to a certain form of governing investment for climate protection, without having the same interest in this form of climate regulation necessarily.

Methods: Document analysis and expert interviews

The analysis of the investment dispositif in climate change governance combines a survey of policy document with expert interviews. The choice of documents and interview partners followed the general objective to identify and describe in detail the discourses and practices that address and thus constitute investment for climate protection.

Though archaeology and genealogy overlap in many ways in the analysis of the climate investment dispositif, it can be said that the archaeological dimension of the research mainly focuses on the survey of policy documents: The analysis of these documents aims at a unit of analysis that Howarth and Glynos (2007) introduce as *logics*.

Building on Laclau, they argue that logics describe the grammar of a regimes of practices or a 'cluster of rules which make some combinations and substitutions possible and exclude others' (Laclau 2000: 76); however, they also entail the conditions that make a practice possible. In short, therefore, the logics concept captures those aspects of a practice or a regime of practices 'which make it tick' (Howarth and Glynos 2007: 135).

Howarth and Glynos use the example of reforms in the British higher education system to illustrate this understanding: they claim that these restructuring processes are generally informed by a neo-liberal governmental rationality, but distinguish between the more specific logics of competition, atomisation, hierarchy, and instrumentalisation that inform the practices of different actors within the universities.

The logics concepts is thus understood here as devising an analytical middle ground between a nominalist perspective that focuses on the formation of concepts on the one hand, and the identification of a general rationality or global strategy within a dispositif on the other hand. Logics, in that sense, describe common patterns of thought or governmental knowledge that are not necessarily reflected (yet) in a common terminology, but specify and differentiate what is contained in governmental rationality.

In the analysis of the climate investment discourse, the objective is therefore to identify widely shared or common understandings with respect to finance and investment for climate protection. Rather than asking for the formation of concepts related to the government of investment, the research focused on the understandings that inform governmental activities in different domains, such as a *lack of finance for climate protection* or the understanding of *risks related to low-carbon investment.*

To identify the logics that drive governmental practices in certain domains, Howarth and Glynos suggest using a process of archaeological bracketing. This process resembles of the methods used in grounded theory research therefore, which moves from concrete to abstract in describing the practices of various actors, assembling the activities that are related to each other that way (Corbin and Strauss 1990).[23]

The choice of documents for this research followed the objective to cover a broad range of actors that contribute to creating investment as an object of government, including governments, IGOs and NGOs. Rather than focusing on a limited number of actors for the important roles they otherwise play in climate change governance, the document analysis seeks to map the diversity of understandings that contribute to the investment challenge in climate governance, and to identify the different roots of these understandings.

The document analysis, in that sense, also contributes to the genealogical dimension of the research: it identifies the discursive and non-discursive practices that aim at investment related to climate protection, and follows them to their points of origin.

---

[23] Grounded theory shares with the logics approach the assumption that the processes of data collection and analysis are interrelated. See also the method of theoretical sampling: Corbin, J. and A. Strauss. 2008. *Basics of Qualitative Research. Techniques and Procedures for De- veloping Grounded Theory*. LA: Sage.

To account for the emergence and change of practices and discourse, the analysis relates the different investment narratives to their historical and institutional contexts. In the case of the climate finance context, for instance, it is important to see how a discursive shift goes along with changes to the institutional context and the main actors involved in the government of climate finance. In the case of the low-carbon narrative, the analysis highlights that low-carbon strategies emerge in a great number of developed and developing countries, and highlights the different perspective on low-carbon growth or development.

The document analysis is complemented by 21 semi-structured expert interviews that were conducted between March 2010 and September 2011 during several international conferences related to climate change and the environment, and with institutions in London and Brussels. While two of these address the significance of the Stern Review and two others focus on the REDD mechanism for forest protection, the majority of the interviews concerns the emergence of the finance and investment challenge in climate governance (see the list of interviews in Annex B).

These expert interviews were in particular important for the genealogical dimension of the analysis, and in filling the gaps that the document analysis left blank. They played an important role in understanding and explaining the emergence of the climate investment discourse, by relating events and developments to each other. The interviews also helped accounting for the understandings, motivations and interests of the actors that play important roles in the emergence of the climate investment discourse, or deploy the government practices that address investment for climate protection.

3   Economic rationality in climate governance

'We are convinced that climate protection makes economic sense, as it would be more expensive in the long term to pay for the damage it causes. It offers companies and national economies that react quickly great opportunities'

(Peter Hoeppe, Munich Re, cf. Mills 2009: 1).

Economic perspectives, interests and strategies have played a crucial role in international climate politics since the very beginning. In 1992, William Cline published his book *The Economics of Global Warming*, and ever since the economic consequences of climate change have been a topic within academia (Cline 1992); business organisations were an important actor and lobby group early in international climate negotiations and managed to prevent stronger regulation in many cases (Levy and Egan 2003); the 1992 UN Framework Convention on Climate Change urges for cost-effectiveness in achieving global climate protection; five years later, the Kyoto Protocol introduces market-based mechanisms to enhance cost-effectiveness; and the development of a global carbon market opened a lucrative playing field for project developers, investment companies and consultancy firms.

Against this background, it seems disputable to claim that the Stern Review *The Economics of Climate Change* (Stern 2006) reframed climate change as an economic problem. However, the study was widely taken up and discussed after its publication in 2006, and apparently made climate change legible in economic terms: The economic consequences of climate change played a much greater role in climate change debate in recent years, and the main argument of the study, that climate protection makes sense from a purely economic point of view, is used widely as an argument for increased efforts for climate protection.

While the Stern Reviews massive repercussion in both mainstream and expert discourse on climate change cannot be questioned, it was certainly not the study alone that changed climate change discourse and governance in recent years. Rather, it is argued here, the Stern Review both reflects and contributes to a growing awareness of the economic dimensions of climate change. This has greatly triggered the importance of the climate issue in general and of economic analyses more in particular, and thus marks a crucial turning point in the climate change discourse.

Two questions arise from this: first, what has caused this massive resonance, as Stern and his team do not present an entirely new perspective on climate change. Rather, they take up and systematise existing economic analyses. And second, whether and how the Stern perspective on climate change effects on climate change governance.

While this he second questions will be addressed throughout this book in various ways, and be taken up more explicitly in the reflection again, this chapter focuses on the way that the Stern Review introduces a new understanding of the economic significance of climate change. It starts with outlining what I will coin the "old" rationality in governing climate change. This first part gives a quick overview of international climate governance from the adoption of the UNFCCC to the publication of the Stern Review. It interprets climate governance since the adoption of the Kyoto Protocol as market-driven climate regime, focusing on the introduction and evolution of carbon markets.

The second and main part of the chapter then introduces the Stern Review as an economic problematisation of climate change. It questions established understandings and procedures in climate governance, expresses difficulties in the way that climate change had been governed until this point, and spells out 'the conditions for formulating answers' (Foucault 2005b) to these problems. It defines the space for addressing climate change in a way that is compatible with a dominant economic rationality, marginalising claims for alternative and sometimes more radical changes to climate change governance.

3.1 The old rationality of governing climate change

This section gives a quick overview of how carbon markets have turned into the centrepiece of climate change regulation after the adoption of the Kyoto Protocol, to then describe the markets and flexibility approach introduced with the Kyoto Protocol as *old* rationality in climate governance.

3.1.1 A short history of carbon markets in climate governance

The first major step in international climate diplomacy was made at the 1992 United Nations Conference on Environment and Development, the so-called Earth Summit, in Rio de Janeiro. Fuelled by hopes that the end of the cold war could be the starting point of a new era of international cooperation, the heads of state and government present at this summit reached a series of agreement on environment and development, including two major international conventions: The Convention on Biological Diversity (CBD), and the United Nations Framework Convention on Climate Change (UNFCCC).

The UNFCCC introduces the Conference of the Parties (COP) as regular meetings of those states that have signed the convention. At the first COP in Berlin in 1995, Parties agreed to negotiate a global treaty on climate change, including binding emission reduction targets and a timetable for achieving these reductions. This treaty was finalised and signed in Kyoto, Japan, in 1997. Until today, the Kyoto Protocol remains the most far-reaching agreement for international cooperation on climate change. Its main achievement is the commitment of industrialised countries to accept binding emission reduction obligations.

This commitment was the result of a series of negotiation sessions between the main polluting countries that often lasted until the early morning hours. Accordingly, the agreement does not simply translate the emission levels and reduction capacity of these countries into binding emission reduction targets, but rather reflects their readiness to go ahead on global climate protection. The European Union, for instance, accepted much higher obligations than the US or Canada, measured against their respective contributions to global emissions.

While some early critics already cautioned that the emission cuts entailed in the Kyoto Protocol would not be sufficient to prevent dangerous global warming, and that the Kyoto Protocol does contain any procedures for realising these emission reductions, the agreement was widely praised as important achievement in tackling climate change.

The following years, however, excelled even the most pessimistic expectations. In 2001, the Bush administration ultimately decided that the US would not ratify the Kyoto Protocol. Given the US' large share in global emission levels, the Protocol could then only come into force in 2005 after the European Union had convinced Russia to ratify the treaty. Nevertheless, even in those countries that have ratified the Protocol, emission levels rose by average rather than to decrease since the adoption of the Kyoto Protocol. Today, the emission targets agreed on in Kyoto have disappeared from the climate change debate almost entirely.

It was thus in another way that the Kyoto Protocol was fundamental in setting a path for the future of global climate governance. The governments present in Kyoto had also agreed to introduce three flexibility mechanisms into the Protocol, to help countries achieving emission reductions in cost-effective ways. *Emission trading schemes* allow trading emission allowances between industrialised countries with binding emission reduction targets. Through *Joint Implementation* and the *Clean Development Mechanism*, these industrialised countries can also finance emission reduction activities in transition economies or developing countries, and count these reductions towards their own obligations.

Carbon markets initially were not meant to become the main tool for climate protection. The Kyoto Protocol states that the emission reductions achieved through the flexibility mechanisms should be additional to domestic efforts only. At the beginning of the Kyoto negotiations, the European Union and developing countries alike had rejected carbon trading altogether. It was a US led group of countries that promoted the flexibility and cost-effectiveness of market solutions, and finally succeeded by refusing to accept any binding obligations for emission cuts without the opportunity to achieve these reductions in flexible ways and beyond their own boarders.

Particularly illuminating for the dynamics in the Kyoto negotiations is the creation of the Clean Development Mechanism (CDM) that has turned into the most important market instrument under the umbrella of the UNFCCC.

Initially, the CDM was not meant to be a market mechanism at all. Its existence goes back to an initiative by developing countries that feared to be excluded from the financial dynamics unleashed by the emissions trading schemes agreed on in the Kyoto Protocol, as only countries with binding emissions targets were meant to participate. Brazil therefore proposed a north-south finance instrument in form of an independent fund for sustainable development, technology transfer and adaptation, filled by payments from industrialised countries that did not achieve their emission reduction targets.

However, many industrialised countries opposed any financial obligations or penalties from a climate treaty. Due to the pressure of the group of countries led by the US, the Brazilian proposal was transformed into an investment instrument that is meant to serve the interests of industrialised and developing countries alike (Oberthür and Ott 1999): CDM investment into emission reductions in developing countries projects should help developed countries to fulfil their mitigation obligations and foster sustainable development in the recipient countries.

But whereas the three flexibility mechanisms were introduced as attachment to the Kyoto Protocol rather than its core part, pollution markets became the centrepiece of global climate governance in the years after Kyoto. One important factor here was the development of the CDM. The World Bank played an important role in financing early CDM activities with public money, and that way demonstrated the viability of CDM investments. As a consequence, private investors increasingly started engaging in the CDM as well. The second important development was the decision of the European Union to create its own Emission Trading Scheme (EU ETS), and to allow participants of the trading system to buy credits from the CDM. That way, the EU ETS became the main (and only commercial) source of demand for CDM credits.

3.1.2 Neoliberal rationality: Markets and flexibility

Taken together, the CDM and the EU ETS became the main engines of a quickly growing global carbon market. The achievements and limits of the Kyoto flexibility mechanisms in contributing to climate protection will be addressed in later chapters. For now, it is important to understand how the creation of these mechanisms is embedded in and contributes to a specific rationality in governing climate change.

The importance of market solutions in achieving climate protection also plays a crucial role in governmentality studies on climate change. While more recent contributions focus on the practices and techniques that transform carbon into a marketable good, it is in particular some of the earlier governmentality studies concerned with climate change that address the rationality involved in market-driven climate governance (Rutherford 2007).

Neoliberal climate governance and the market superiority narrative

These earlier studies distinguish two governmentalities as relevant in addressing climate change: green governmentality/biopower and advanced liberal government/ecological modernisation. While the former was crucial in rendering climate change governable, the latter explains the choice of market solutions in climate governance.

Luke (1996) introduced the *green governmentality* concept to describe the role of environmental studies in bringing about a specific power/knowledge formation in addressing (global) environmental problems. Environmental science as a new academic discipline, in that sense, creates both the knowledge and the professionals required for the management of the global environment. This way, the entire planet 'can be reduced [...] to a complex system of interrelated "natural resource systems," whose constituent ecological processes are left for humanity to operate – efficiently or inefficiently – as the geo-powers of one vast terrestrial infrastructure' (Luke 1996: 2).

The new eco-knowledges thus help extending government control – or biopower – to the entire planet (Bäckstrand and Lövbrand 2006). Green knowledge and expertise make visible and thus manageable global environmental problems in a specific way; they bring about 'an elitist and totalizing discourse that effectively marginalizes alternative understandings of the natural world' (ibid. 55). Luke thus defines green governmentality as 'an instance of reinforcing the power of the administrative state in the name of responsible stewardship of nature' (Luke 1999a: 129). The green governmentality perspective has also been applied to the way that climate science made climate change visible and thus governable as the regulation of global emission levels.

Luke's understanding of green governmentality also links the management of the global environment to the maintenance and expansion of capitalist growth. The eco-knowledges he addresses aim 'at generating geo-power from the more rational insertion of natural and artificial bodies into the machinery of global production [...] in diffuse projects of ecological modernization' (Luke 1996: 3: 105, cf. Oels.).

Oels (2005) and Bäckstrand/Lövbrand (2006), to the contrary, suggest distinguishing green governmentality from ecological modernisation strategies. The latter are embedded within neoliberal governmentality and foreground the economic dimension of environmental or climate protection. They also agree that this should not be understood as a historical shift from one governmentality to another, at least not completely: Whereas Oels (2005: 32) suggests that the apparatuses of biopower 'may have reconfigured themselves within economic terms of reference in order to secure their existence', Bäckstrand and Lövbrand (2006) claim that both are combined in climate governance and enabled the creation of the Kyoto institutions.

Bäckstrand and Lövbrand draw on ecological modernisation in the way that Hajer (1995) uses the concept, to describe the growing importance of an economic perspective on climate change governance. The ecological modernisation discourse emerges with the Brundtland report *Our Common Future* in the 1980s. Ecological modernisation strategies aim at aligning economic growth with resource conservation, through 'a gradual transformation of the state and market to promote green regulation, technology, investment and trade' (ibid. 53).

Following others, Bäckstrand and Lövbrand distinguish a weak version of ecological modernisation as 'a technocratic and neo-liberal economic discourse that does not involve any fundamental rethinking of societal institutions', from a strong and more critical version that 'entails greater institutional reflexivity, democratization of environmental policy and a focus on the justice dimension of environmental problems' (ibid. 53).

The central idea of weak ecological modernisation is to use markets, or market instruments, to promote cost-effectiveness through flexibility in environmental protection. Ecological modernisation, in that sense, replaces or transforms top-down approaches to environmental governance that were increasingly perceived as not flexible enough and too expensive.

In that sense, the emergence of weak ecological modernisation as the dominant paradigm of environmental governance in recent decades has been understood as embedded within the rise of neoliberalism, or a change of governmentalities from biopower to advanced liberal government (Oels 2005). 'The hierarchical legislative system often involved complicated administrative procedures which became problematic in a period in which deregulation became widely accepted as one of the goals of government' (Hajer 1997: 27)

The weak variant of ecological modernization advocates free market settings and limited government interventions as a way of incentivising technological innovation that will solve environmental problems in the most cost-effective way. It also points to the economic opportunities of addressing environmental problems, and in that sense 'reconceptualises the ecological crises as an opportunity for innovation and reinvention of the capitalist system' (Oels 2005: 22). Altogether, environmental problems cease to be a moral issue and become subject to cost-benefit-analyses and market-oriented instruments.

The diagnosis of neoliberal climate governance usually coheres around the Kyoto institutions and the creation of the EU ETS. 'Public policy over the last 25 years has been dominated by neoliberal ideology which has driven solutions to emerging social, political and economic problems. Given this, it is not surprising that emissions trading schemes founded on the core tenets of neoliberalism have emerged as the prevailing response to climate change by developed countries' (Bailey and Maresh 2009; Andrew, Kaidonis et al. 2010).

The Kyoto Protocol with its focus on cost-effectiveness and market flexibility reflects the prominence of the weak ecological modernisation discourse in climate governance. The CDM in particular epitomises the win-win logic of ecological modernisation, combining sustainable development investment in the South with low-cost mitigation for polluters in the North.

To specify the role of neoliberal economic rationality in environmental and climate governance, Oels draws on the notion of *advanced liberal government* that Rose (1993) introduced to describe a specific and more recent type of neoliberalism. 'Advanced liberal government introduces the market as organizing principle for all types of social organization including the state. Advanced liberal government employs market forces to guarantee freedom from excessive state intervention and bureaucracy' (Oels 2005: 191)

The Kyoto Protocol, she argues, is embedded in advanced liberal government rationality: It sees states as the actors responsible for finding solutions to climate change, and describes sees states as rational and calculative actors that pursue the least-cost solutions to climate change. The failure of states to combat climate change in a cost-effective way has thus reduced the scope for state intervention on behalf of climate change: 'In advanced liberal government, the range of available policy instruments is more or less limited to market-based solutions that spur technological innovation and economic growth' (ibid. 201).

Trading carbon, making things the same

However, we have to further clarify what is specific about a carbon market approach to climate protection, as carbon markets are not the only possible form of market-oriented climate governance. Environmental taxes as another popular option for reducing carbon emission are today often portrayed as more regulatory approach. Initially, however, both pollution markets and environmental taxes were introduced as market based solutions to environmental problems: both work through a price signal and market choice.

It is thus important to specify how carbon markets are different from other forms of market-oriented climate governance. The Kyoto approach to climate protection aims at achieving cost-effectiveness in climate protection through trading carbon emissions in a global market. This requires giving carbon a monetary value by turning it into a commodity.

Before the creation of carbon markets, the emission of GHG was largely free and without restriction. Emission trading schemes like the EU ETS therefore assign a limited number of emission rights or certificates to carbon emitters and turn these emission rights, and thus emissions themselves, into a scarce good; through this scarcity, emission permits gain an economic value for polluters and achieve a price in carbon markets.

The fundamental precondition for establishing a carbon market, then, is to make various objects and activities comparable and commensurable with respect to their contribution to emission reductions: this then allows for trading these otherwise different objects and activities in a common market.

The creation of carbon markets has therefore also been described as 'making things the same' (MacKenzie 2009: 441). The commodification of GHG and the trade in carbon markets makes very different objects comparable through a common metric. The certificates for carbon reductions that are issued as part of the CDM or the EU ETS all carry the same value, 'despite the potentially very different material circumstances (a forest compared to a wind farm) in which they were produced. In this way, carbon is individuated (separated from its supporting context) involving a discursive and practical cut into the world in order to name discrete chunks of reality that are deemed to be socially useful' (Bumpus and Liverman 2007: 136).[24]

---

[24] One very specific challenge in that sense that caused trouble for scientists within the IPCC and beyond was to agree on the global warming potential of different GHGs, and thus on the relation with

The process of commodification means that the conditions and circumstances of the production of the commodity are not visible in the abstract form of emission certificates, and therefore play no role in the decisions of market participants. This abstraction from the material circumstances of a commodity is no random side effect, but an essential feature of commodification and indispensable for the functioning of markets.

Robertson illustrates the relationship between abstraction and commodification in the case of wetland mitigation banking in the USA (Robertson 2004). The program allows the destruction of a wetland, for instance to create building land, if an adequate substitute is formed elsewhere. The certificates issued for the creation of wetlands can also be traded in markets.

Conflicts arose, however, over the appropriate valuation of wetlands. Size alone does not reflect the various functions a wetland can have, such as plant species diversity or water cycles; trading wetland certificates on the basis of wetland size exclusively would thus make these functions invisible and thus irrelevant for the decisions of market participants. Representing these functions in the certificates, to the contrary, made certificates less commensurable, and therefore inappropriate for market trade.

The creation of a global carbon market makes objects commensurable that are much more diverse then the different types of wetlands in Robertson's example. Within the EU ETS, the same metric is applied to different industries and forms of energy production. More important in that respect, however, was the decision to link the EU ETS and the CDM. While in the case of trading carbon in the ETS, it is only forms of GHG emissions that are made commensurable; the CDM also adds the non-emission of GHG to the list of available options. CDM projects include the destruction of industrial gases, the creation of wind or solar power plants, and the planting of trees: with respect to their market value, these activities are all equalised on grounds of the carbon emissions they help to prevent or reduce.

Summing up, we can describe the approach to climate protection represented through the Kyoto Protocol and the creation of carbon markets as the pursuit of cost-effectiveness in climate governance through market flexibility. The decision over which climate protection activities to pursue and which investments to take is left to market participants. Mitigation activities are exclusively chosen on grounds of their contribution to reducing carbon emission.

---

which the reduction of these gases is traded in emission markets: The industry gas HFC 23, for instance, has a warming potential that is 11,700 times higher than that of CO_2, so that reducing one tonne of HFC 23 is equivalent to reducing 11,700 tonnes of CO_2, and carries the same market value accordingly.

Other characteristics, such as the broader environmental or social consequences of emission reduction activities, are not visible to market participants and thus play no role in funding decisions.

Returning to the understanding that different economic problematisations of climate change are possible, we can understand the choice of market mechanisms in climate governance (and beyond) as the expression of a particular economic rationality that focuses on safeguarding economies and markets from the consequences of interventions on behalf of climate change. Oels, in that sense, suggests that 'advanced liberal government renders economic growth as the entity to be secured from excessive climate protection costs' (Oels 2005: 201).

The rest of the book will address the question whether and how we can see changes to this rationality in climate governance in recent years. The rest of this chapter is concerned with how the Stern Review challenges some of the understandings that let to the creation of carbon markets. The next two chapters highlight that the climate investment discourse suggests that investment decisions related to climate protection must be taken more explicitly, and thus contributes to unmaking things the same in climate governance.

3.2 Stern and the economics of climate change: changing perspectives

The Stern Review plays an important role in challenging the rationality in climate governance established with carbon markets. Highlighting both economic risks of climate change and the economic opportunities of enhanced climate protection, it raises the awareness for the economic significance of addressing climate change, and problematises established forms of governing the climate.

3.2.1 Reframing climate change as an economic problem

The Stern Review has played a major role in pushing climate change to the centre of the political agenda. The Stern Review was published in 2006 by the British government and resonated heavily in mainstream and expert media. The massive global uptake of the study came as a surprise to the authors and other experts alike.

Mike Hulme, a prominent academic concerned with climate change in the UK, suggested at the time of its publication that 'The Stern Review resonates well within the domestic political landscape in Britain, but perhaps nowhere else. [...] But how effective will it be and what difference will it make? In ten years time, will we be able to look back and analyse a pre-Stern and post-Stern discourse about climate change, or see 2006 as marking a break-point in climate policy? I suspect not' (Hulme 2007b: 1).

Hulme was definitely wrong in expecting little resonance of the Stern Review beyond the UK. And one could also argue that it makes sense to speak of a pre-Stern and a post-Stern climate discourse—as long as this does not imply that the Review caused this change on its own. With the publication of the report, awareness for the economic dimension of climate change was spread beyond the inner circle of economists and climate professionals (Egner 2007). Policy makers and scientists alike rely heavily on his work to add legitimacy to their arguments and calls for enhanced climate protection efforts, and Nicholas Stern, not particularly a climate expert until then, is today seen as 'the global authority on climate change' (The Guardian).

The question remains, however, what caused this massive resonance, as Stern and his team do by no means present radical new insights into the climate problem. Rather, they review, reflect, and reconsider existing economic analyses on climate change, and use their calculations for a strong economic argument for climate protection.

The remainder of this chapter therefore demonstrates that the Stern Review reinforces a changing understanding of the climate challenge that focuses on the economic risks and opportunities related to climate change, and is taken as the starting point for new strategies that align climate change with economic objectives. To adequately understand these changes, we must distinguish between the study's main message that was widely taken up in the public debate and heatedly discussed in expert circles; and a new approach to combating climate change that the Stern Review helps to establish, and which subsequently informs much of the (global) climate discourse.

The main message: political, not scientific, argument

Stern and his team compare the global macroeconomic costs of climate change, with and without mitigation policies, for time horizons of 50, 100 and 200 years (Stern 2006). The core message is that the economic consequences of uncontrolled climate change far outweigh the costs of limiting climate change to an acceptable degree; mitigating climate change, therefore, is perfectly rational from an economic point of view.

It is this message that was strongly taken up and almost entirely welcomed in public debate on climate change. After the publication of the Stern Review, a growing number of similar approaches took on the enquiry into the economics of climate change for particular sectors or world regions. Likewise, a growing number of publications and research projects analyse the costs and benefits of adaptation policies (see for example European Environment Agency 2007).[25]

The great number of similar studies published in the aftermath, in many cases with explicit reference to the *Economics of Climate Change*, shows that the Stern perspective on climate change is seen as a fruitful approach, and that 'the economic perspective has shown to be helpful in broadening the climate discourse' (Luks 2008: 187, own translation). It may be for this reason that even critics admit Stern is 'right for the wrong reasons' (Weitzmann 2007: 723).

However, the main message of the Stern Review is a political rather then a scientific argument, and of limited value for concrete political decisions. The constraints to the – scientific – relevance of the approach become apparent with a look at the critical debate among economists that followed the publication of the review (Wolf 2009).

The debate among economists considers the generation of concrete numbers, the chosen methodology, and the assumptions taken therein. While this debate does not question the economic perspective on climate change in general, the discussion of methods and parameters highlights the degree to which the results of the Stern Review (and comparable studies alike) depend on the decisions taken by scientists. This is due to a number of reasons:

- Uncertainties surrounding the consequences of climate change are reproduced in economic scenario building. Translating damages into economic terms and numbers requires an additional intervention – or translation – by scientists, and hence accelerates the ambiguity of estimates.
- The costs of economic consequences and mitigation efforts alike do not only depend on environmental changes. Economic and technological development is an important factor that has to be taken into consideration and requires additional assumptions.[26]
- Modelling costs and benefits demands the choice of a normative standpoint. Intergenerational justice, distribution issues, or the value of natural resources are issue that require consideration and judgement here.
- All this suggests that political organisation and power relations play an important role in defining mitigation and adaptation needs and the choice of political strategies, and thus are a factor in calculating the costs of climate

---

[25] However, so far there is little evidence that benefits of adaptation policies will outnumber the expected costs, and calculations are even more 'highly sensitive to assumptions and uncertainties' here (Dyszynski 2008).

[26] Leaving these factors unconsidered implies the assumption that there will be no technological development.

change and its mitigation as well; this is another influence factor that is hardly reflected in economic modelling.

Based on the calculations in an earlier IPCC report, Rayner (2000: 277) suggests that economic estimates of climate change damages carry an error rationality of at least 50 per cent, and thus concludes that these forecasts remain, regardless of technological and scientific developments, 'no more or less than a consensual, subjective judgment'.

Debates on the appropriate discount rate

While many issues were raised regarding the methodology chosen in the Stern Review, the debate culminated around the choice of the discount rate (Barker 2008).[27] This factor reflects a standard operation in economic modelling and investment calculation. Put very simplistically, economists act on the assumption that the value of investments, just as the value of money in general, diminishes over time, as future generations are expected to be wealthier through economic growth. Consequently, financing a certain measure of mitigation or adaptation requires a smaller share of GDP when realised in the future instead of today. The resulting discounting of present investment is what economists call *time prefer-ence*. Furthermore, many studies add *pure time preference* as second discount factor that reflects the argument that people prefer to possess or consume goods earlier rather than later (Stern 2006).

Stern and his team choose 1.2 per cent as *time discount rate*, which is sig-nificantly lower than in comparable studies, and refuse to add a *pure time dis-count rate* for ethical reasons: 'Attaching little weight to the future, simply be-cause it is in the future ("pure time discounting"), would produce low estimates of cost – but if you care little for the future you will not wish to take action on climate change' (Stern 2006: 143). The Stern Review thus differs significantly from the 'conventional value' of six per cent in comparable studies (Weitzmann 2007).[28] As a result, today's climate protection efforts have a higher value in the Stern Review than in other models (Barker 2008).

It is these assumptions that were criticised most heavily by other econo-mists. A higher rate of time preference, they insist, would have resulted in far

---

[27] For an overview of critics and comments to the Stern Review, see Yohe and Tol 2008. For more details on discounting in the Stern Review, see Spash 2007; on discounting in general, see Price 1993.

[28] Taken together, the discount rate in the Stern Review sums up to 1.4 per cent, a value significantly lower than in comparable studies.

less clear conclusions about the costs and benefits of climate protection. Arrow (2007) counters this argument. He agrees that Stern's results differ considerably from other studies but suggests that even choosing a much higher discount rate (up to 8.5 per cent following his calculations), climate protection would still be economically rational (Arrow 2007). Yohe and Tol (2008), among others, therefore do not criticise Stern and his team for choosing a too high or too low discount rate, but rather for opting for one single value, instead of working with different scenarios.

Likewise, they criticise that the report underestimates the likely costs of global mitigation programmes, an argument that is once again countered by Schneider (2008), as 'it is unacceptable to compare future costs to the present scale of the economy, [...] since projected growth rates of the economy swamp all mitigation and adaptation costs typically found in the literature' (Schneider 2008: 3).

Summing up these arguments, it becomes obvious that the choice of a distinct time discount rate cannot be justified through scientific reasoning. Rather, it reflects particular world-views or beliefs, and normative assumptions, as Stern and his team point out: 'Modelling over many decades, regions and possible outcomes demands that we make distributional and ethical judgements systematically and explicitly' (Stern 2006: 143).

It is important parameters in economic modelling which cannot be calculated scientifically, yet must be predefined or estimated. Without taking particular normative assumptions, or by taking others, the recommendations of the Stern Review could have been completely different. 'The high discount rates favored by many economists seem to justify doing very little for now; Stern's low discount rate, applied to the same data, endorses doing much more, much sooner' (Ackermann 2007: 16). Correspondingly, some of its critics challenge Stern's conclusion, 'because there remains substantial uncertainty about the extent of the costs of global climate change and because these costs will be incurred far in the future' (Arrow 2007).

This makes Yohe und Tol (2008: 237) ask whether it is necessary to rely on modelling the costs and benefits of climate protection at all: 'The point we want to underline here is that these economic arguments can be made without resorting to dodgy modelling or peculiar assumptions'.

The main message of the Stern Review, accordingly, was not a purely scientific, but rather a political argument—a point of view that Stern himself supports three years later in his *Blueprint for a safer planet* book. Here, he suggests that 'as a simple way of communicating, this formal modelling-based approach had and continues to have some usefulness. [...] It is crucial to emphasise, however, that the precise quantitative results of such models should not be taken too serious' (Stern 2009a: 94).

Global analysis versus national strategies

Additionally, the Stern Review's main argument is of limited value from a policy perspective, as it carries an intrinsic contradiction if regarded from a strategic point of view. Important climate policy decisions are taken at the national level—this holds true not least for commitments to binding emissions targets. A national calculus, however, cannot be derived from the global comparison of costs and benefits.

Even taking the argument for granted that global climate protection is economically rational in the middle and long-term, the situation can be very different from a national policy perspective. Emissions reductions of an individual country do not limit the effects of climate change for that same country, and it is by no means certain that the effects of climate change will be economically negative in every case.

The expected damages through climate change, as much as the expected costs of protection measures, differ sharply between countries and regions. Comparable to the mean change in global temperature which is not the relevant figure for individual countries, the global loss in economic value is 'demonstrably unimportant in leading to actual impacts' (Rotmans and Dowlatabadi 1998: 311). Global costs are no contradiction to local or national gains.

For a number of countries, it is not yet possible to say whether climate change will cause economic losses or gains, for at least two reasons. Rising temperatures create new opportunities for some economic sectors, and adaptation policies in general can be seen as an economic stimulus. In an interview with the German news magazine *Der Spiegel* in August 2010, Hans Joachim Schellnhuber, head of the Potsdam Institute for Climate Impact Research and advisor to German chancellor Angela Merkel, said 'it is necessary to pay attention to the opportunities and advantages for some world regions as well. At least temporarily, there will be winners as well, in particular in the northern hemisphere'.[29]

What Schellnhuber does not even consider in this argument are the economic opportunities and benefits from mitigation and adaptation: building climate resilient societies and advancing low-carbon growth can have stimulating effects on economies, as the Green New Deal debate suggests; the growing de-

---

[29] 'Tritt in den Hintern'. Interview in: Der Spiegel, 33/2010, August 16, 2010 (own translation).

mand for climate proof technologies creates new export markets, and low-carbon technologies are already an important sector in several countries.

Consequently, comparing costs of national mitigation efforts with economic effects of climate change at the national scale could end up in a plea for skipping emissions reductions. The positive effects of mitigation efforts, therefore, cannot be judged from an economic perspective alone. Rather, their influence is politically mediated, in that they give incentives to other countries to mitigate emissions as well.

Even when the motivation to climate protection is explained for within a global perspective (rising from solidarity, responsibility, the interest in greater influence in world politics, or the perception of security threats through climate change), the contribution of a global cost-benefit analysis is rather limited. Stern and his team can only inform us that the economic effects of emission reductions in the (very) long-term are possibly positive.

The Stern Review therefore is, as one of its authors suggest, 'a standard piece of normative welfare economics, which asks the question: assuming a benevolent global player, what would you do?'.[30]

The answer to this question could be that a benevolent global player, should it exist, would probably make a strong effort to limit climate change, but turn down economic arguments for much stronger normative motivations. This sheds light on the grounding of the report's argument: While the Stern Review claims to make an economic argument, it totally depends on normative considerations. The economic case for climate protection is justified through non-economic reasoning, and policy makers have to take the general decision for climate protection without relying on cost-benefit considerations.

3.2.2 Turning climate governance into an economic challenge

Nonetheless, the general argument of the Stern Review had great effect on climate policy debates. Providing the economic rationale for climate protection, it can be said to have supported those who wanted to make a strong case for emissions cuts, and strengthened their power within governments.[31]

---

[30] Interview, Stern Review co-author.
[31] Interview, Stern Review co-author.

The economic case for climate protection

This is by no means accidental. Certainly, the report fitted well into the political landscape, as the G8 meeting in Gleneagles in 2005 had already pushed the climate change issue up the political agenda, and the upcoming fourth assessment report by the IPCC forewarned that the consequences of climate change would be much more severe than expected so far.

But it was also the institutional context of the Review's publication that increased its importance. Gordon Brown who at that time was Chancellor of the Exchequer in Great Britain commissioned the Stern Review. He aimed at putting as much authority behind the *Economics of Climate Change* as possible, by making it an official government report written by a former World Bank chief economist.

The objective was to make a strong case for the economic dimensions of climate change, and to bring finance ministries into the game. The assumption of the British government, as one of the authors explains, was that climate change would only rise up the international political agenda if it could receive a push to its credibility as an economic issue. In that sense, it can be argued that the Stern Review 'is kind of a grant geopolitical plan the government had to increase the profile of climate change and to persuade other governments to give it attention [...]. It wouldn't have received the same attention, it wouldn't have been so important had it been undertaken by DEFRA.[32] [...] So I think it was a political choice'.[33]

Bringing together political authority, economic expertise and the input of a great number of research institutes within the UK and beyond, the report therefore 'has cache in three of the circles of institutionalised power – politics, economics and science. [...] *The Stern Review* is a megaphone voice from London speaking into the clouds of hesitation, deviation and prevarication which surround current international negotiations' (Hulme 2007b: 1).

Gordon Brown certainly achieved the first part of his objective, as the attention the report gained far exceeded the authors' expectations. Of course, critical voices were raised as well.[34] Like others, Ruth Lea, Director of the Centre for Policy Studies, questions the feasibility of the modelling efforts in the Stern Review and with it the desirability of mitigation policies, as 'the climate system

---

[32] The UK Department for Environment, Food and Rural Affairs (DEFRA) was then the government body responsible for climate change issues.
[33] Interview, Stern Review co-author.
[34] The Wikipedia entry on the Stern Review lists a great number of articles and comments on the Stern Review: http://en.wikipedia.org/wiki/Stern_Review. A BBC webpage also collects (mainly positive) reactions: http://news.bbc.co.uk/1/hi/business/6098612.stm.

is far too complex for modest reductions in one of the thousands of factors involved in climate change (i.e., carbon emissions) to have a predictable effect in magnitude, or even direction'.[35]

Miles Templeman of the Institute of Directors, the UK's largest business lobby organisation, cautioned that any climate policy that does not include other main polluters 'would be bad for business, bad for the economy and ultimately bad for our climate'.[36] Other business organisations and their organic intellectuals likewise cautioned against climate policies that harm economic performance and growth. An article in the Reason magazine, in that sense, suggests 'that if one wants to help future generations deal with climate change, the best policies would be those that encourage economic growth'.[37]

Environmental writer Bjørn Lomborg, who had already gained some reputation as a climate sceptic and critic of climate protection efforts, found that 'despite using many good references, the Stern Review on the Economics of Climate Change is selective and its conclusion flawed. Its fear-mongering arguments have been sensationalised, which is ultimately only likely to make the world worse off'.[38]

But overwhelmingly, the Review was welcomed and praised. The British government published, simultaneously to the report itself, a list of those who emphasize the importance of the report, including several Nobel Prize economists and representatives from the World Bank, the International Energy Agency, and the UN Secretary General. Many governments, businesses and civil society organisations also welcomed the message that 'we must act now', or, as Simon Retallack of the think tank IPPR put it: 'This removes the last refuge of the "do-nothing" approach on climate change, particularly in the US'.[39] Many business and investor organisations welcomed that the report emphasizes the compatibility of climate protection and economic growth, and described the need for enhancing the scope of markets as important business opportunity.

---

[35] Ruth Lea: 'Just another excuse for higher taxes. The Telegraph, October 31, 2006: http://www.telegraph.co.uk/comment/personal-view/3633767/Just-another-excuse-for-higher-taxes.html..

[36] The Telegraph, October 30, 2006: http://www.telegraph.co.uk/finance/2949852/Amber-alert-over-green-taxes.html.

[37] Roland Bailey: 'Stern Measures. Adverting climate change is surprisingly affordable. Or is it?' Reason, November 3, 2006: http://reason.com/archives/2006/11/03/stern-measures.

[38] Bjorn Lomborg: 'Stern Review: The Dodgy Numbers behind the Latest Warming Scare'. Wall Street Journal, Online Edition, November 2, 2006.

[39] BBC online: Expert reaction to climate change: http://news.bbc.co.uk/1/hi/business/6098612.stm.

Closing the gap between economic theory and climate change reality

It is important to understand the difference between the highly critical debate among economists and the broad appraisal of the Stern Review in policy debates; 'the Stern Review has been accepted by governments and the public as mainstream and consensus economic thinking, when in critical respects it represents a radical departure from a traditional deterministic treatment and its messages for economic policy' (Barker 2008: 175). This is consistent with the observation of one of the Stern authors that the report 'triggers overwhelmingly positive associations within policy circles, business, and civil society', whereas among academics, especially among economists, being an author of the Stern Review means "to have written controversial things"'.[40]

One possible explanation for this seeming paradox is a growing misfit, in the years before the Stern Review, between the evidence for the huge and potentially catastrophic consequences of climate change, and the still prominent argument in mainstream economics that the economic consequences of climate change would be comparatively small and compensated for by economic growth. Nicholas Stern addresses this misfit in his 2009 *Blueprint* book. Here, he laments that recommendations for a conservative approach to climate protection resulting from traditional economic modelling were widely perceived as *the* economic perspective on climate change, and that this has played into the hands of those who opposed ambitious efforts to stop global warming (Stern 2009).

The conclusions of the Stern Review thus matched much better with the concerns of many governments, and the urgent need to formulate more ambitious climate protection strategies.[41]

Two features in particular distinguish Stern's approach from many other economists, and help not only to explain the choice of a very different discount rate, but also play an important role in policy debates. First, the Stern Review gives more emphasis to existing and lingering uncertainties and the risk of potentially catastrophic consequences of climate change. This is explicitly factored into the discount rate, but informs the report in a much broader sense as well. And second, the Stern team pays more attention to the opportunities that arise from climate change and climate protection and call for a more active role of governments and policy-making to realise these opportuntities.

---

[40] Interview, Stern Review co-author.
[41] Interview, Stern Review co-author.

Risk, uncertainty, and the limits to economic modelling

Barker (2008) describes the difference between the Stern Review and a large share of the economic literature on climate change as part of a wider shift from *old* to *new* economics: Old or traditional neoclassical economics, in that sense, is characterised by an emphasis on rationality in the form of utility maximisation, the neglecting of strong kinds of uncertainty, and the use of traditional cost-benefit analysis (CBA) that compares the benefits of abatement policies against the costs of climate change (Beinhocker 2006 Maréchal 2007).

In particular the use of traditional forms of CBA that is seen as inadequate and misleading: CBA 'does not yield transparent or objective evaluations of benefits; rather, it renders the discussion of benefits obscurely technical, excluding all but specialists from participation. At the same time, political debate continues behind the veil of technicalities, as rival experts battle over esoteric valuation problems' (Ackermann 2004).

One particularly important difference is also that fundamental uncertainty plays no role in the modelling efforts that use the tool box of traditional economic theory; rather, current ignorance of the potential climate change consequences is taken as an argument to wait and do nothing here, or to wait for more accurate information that allows to calculate the marginal damage costs of emitting greenhouse gases.

Tol (2005) for example suggests that 'climate change impacts may be very uncertain', but still claims that it is unlikely that the marginal damage costs of carbon dioxide emissions exceed the threshold of US 14 per ton CO2, and suggests that they 'are likely to be substantially smaller than that' (Tol 2005: 2073). Updating his study two years later he calculates a marginal damage cost of US 4 for every ton of CO2. Drawing on these numbers, Nordhaus concludes that 'the Gore and Stern proposals [...] are more costly than [doing] nothing' (Nordhaus 2007: 201).

The Stern Review, to the contrary, factors the potentially very strong consequences of climate change into its estimates explicitly, in the form of climate sensitivity (the possibility that the climate system reacts more strongly to increasing levels of atmospheric CO2 concentrations than expected), and the possibility of an abrupt climate catastrophe (Ackermann 2007). Taking into account risk and uncertainty this way strengthens the argument for mitigation. The Stern Review is about the economics of managing very large risks, 'uncertainty is an argument for a more, not less, demanding goal' (Stern 2006: xvii).

Yohe and Tol describe this as cherry picking of the more severe forecasts of climate change consequences for Stern's analysis. While many economists likewise criticise the way in which the Review treats risk and uncertainty to make an argument for climate protection (Weitzmann 2007), some critics welcome the emphasis on risks and acknowledge the Review's sophisticated methodology: '[...] in general the risks and uncertainties of abatement are much less than those of the damages. If decision-makers are risk averse, or wish to follow the Precautionary Principle, the fact that abatement costs are less uncertain is likely to justify action involving higher costs than what otherwise would be the case' (Barker 2008: 183).

Perspectives thus vary widely on what risks are and how they should be treated in economic modelling, or whether large risks and uncertainties even limit the scope of cost-benefit analysis. 'If the scenario is for the end of the world (as we know it) in the near future, it would obviously make little sense to talk about costs and benefits. [...] If total catastrophe is feasible, then [...] the only rational action would appear to be to stop global warming immediately' (Pearce 2003: 365). But: 'No one appears to argue that catastrophe consists of the destruction of the entire human race. Rather, the kinds of events that are discussed are the melting of the West Antarctic ice sheet or reversal of the gulf stream. Thus it seems more correct to refer to extremely high marginal damages occurring with some unknown probability, rather than marginal damages being infinite' (ibid.).

Weitzmann (2007), to the contrary, accepts that the existence of large risks sets limits to economic modelling: '[I]n rare situations like climate change [...] we may be deluding ourselves and others with misplaced concreteness [...]. Perhaps in the end the climate-change economist can help most by not presenting a cost-benefit estimate for what is inherently a fat-tailed situation with potentially unlimited downside exposure as if it is accurate and objective' (Weitzmann 2007: 18).

Opportunities, markets, and the role of government

The Stern Review turns the complexity of climate change modelling, which might be seen as limiting the role of cost calculations in formulating response strategies, into an argument for strong and early action. It is this argument that was broadly welcomed in policy debates, though these debates would usually take little notice of the arguments that economists exchange on the appropriateness and scientific validity of the Stern Review.

A second issue that distinguishes the Stern Review from much of the economics literature on climate change (until that time) is its strong emphasis on the opportunities that arise from climate change, despite or even because of the need for climate protection. This is contained in the main message that financing climate protection today is less costly than coping with the uncontrolled consequences of climate change in the future. The Stern Review, in that sense, concludes that '[t]ackling climate change is the pro-growth strategy for the longer term, and it can be done in a way that does not cap the aspirations for growth of rich or poor countries' (Stern, 2006: viii).

The argument is spelled out in more detail in Stern's following publications (Stern 2008; Stern 2009a). Here, Stern promises a new wave of investment and technology development if the appropriate regulation is in place. Climate change presents opportunities not only for low-carbon technology developers, but also for banks and investment companies that can benefit from managing the capital flows to low-carbon industries. This emphasis on economic opportunities explains much of the positive response to the Stern Review, and Stern is aware of the strategic dimension of this perspective: 'Demonstrating these arguments country-by-country is critical to building support for action' (Stern 2008: 41).

The emphasis on government support, ones again, counters the assumptions of (traditional) economics that technological change is exogenous to the political and economic system and that economic instruments, such as taxes or carbon prices, cannot push this transformation; it would thus be appropriate to 'wait rather than act now because the costs of mitigation will fall as a result of technological change' (Barker 2008: 179).

The new type of economics that Stern and his team advance, to the contrary, emphasizes the role of institutions in providing incentives for markets and investment. Stimulating technological change through political intervention can thus reduce the overall costs of transformation. It is this way in which Stern problematises the role of state intervention for technological change, and thus the opportunity to align climate protection with economic growth and prosperity. We will later see that this argument plays an important role in the climate investment discourse as well.

3.2.3   The limits to economic rationalisation

The Stern Review describes a win-win perspective on climate change but likewise emphasises that the required transformation will entail economic and political costs. 'These costs must be acknowledged and managed, not dismissed' (Stern 2009: 37). Overall, however, the benefits from climate protection and correcting market failures through regulation and incentives are expected to clearly outnumber the economic losses and dangers (Stern 2009).

The way in which the Stern Review reframes climate change as an economic problem, however, goes beyond the correction of a few market failures. In a more fundamental way, it allows expressing the complexity of climate change in monetary terms, and thus makes climate challenge legible as economic problem.

Despite all the praise that the Stern Review received for its balanced approach, this framing is not free from contradictions and critical issues. Spash (2007) provides a comprehensive overview of these aspects. He emphasises that Stern and his team address important issues that many other economists ignore, including strong uncertainty, plural values, non-utilitarian ethics, distributional inequity, and the treatment of future generations. Stern, in that sense, argues that it is important 'to consider a broader range of ethical arguments and frameworks than is standard in economics [...]' (Stern 2006: 31).

However, Spash asks: 'How then can this report, acknowledging so many of those aspects of climate change that render CBA an unsuitable tool for generating policy recommendations, go ahead to conduct a global CBA and make policy recommendations?' (Spash 2007: 706). His answer is that in the Stern Review important issues are 'suppressed and sidelined in a careful and methodological manner' (ibid.): the modelling techniques applied by Stern and his team fundamentally reduce the complexity of climate change and its consequences. This is also the precondition for making economic growth the central success criterion for climate protection strategies.

Additional to the issue of discounting that was addressed above, modelling the costs of climate change and climate protection raises the more fundamental challenge to find a common denominator for the very different issues that are at stake. It involves, on the one hand, to render very different things comparable, like the consequences of income loss in rich and poor countries; and on the other hand, to express the many non-monetary consequences of climate change in monetary terms, including the value of ecosystems and livelihoods.

The challenge for Stern and his team is thus even greater than it was for a study led by Robert Costanza in the 1990 that calculated the *value of the world's ecosystem services and natural capital* (Costanza, d'Arge et al. 1997). The study raised a lot of interest in the academic world for its broad and ambitious approach. However, while Costanza and his colleagues aimed at estimating the *present* value of natural capital on a global scale, the Stern Review additionally contains estimates for the value of *changes* to or *losses* of ecosystems.

Nevertheless, the discussion around the Costanza study provides interesting insights into the limits of modelling the economic value of global ecosystems, and in particular into the consequences that are drawn from this. A great number of scholars raised concern with Costanza's approach, ranging from general objections to more specific methodological issues. But most critics recommended to intensify modelling efforts rather than to dismiss the approach altogether due to ubiquitous constraints (Daly 1998; Herendeen 1998). Often, they acknowledged a positive effect on the environmental discourse: 'We noted the advantages of speaking in the dominant economics language. [...] No doubt, even though we questioned whether this article would enrich anyone's personal understanding of the value of the earth's ecosystems, we expected it would increase the effectiveness of environmental arguments and thereby influence environmental policy' (Norgaard and Bode 1998: 38).

The authors of the Stern Review are likewise critical about modelling the complexity of climate change and its consequences in economic terms: 'we have conceptual, ethical and practical reservations about how non-market impacts should be included, although there is no doubt they are important. [...] Economists have developed a range of techniques for calculating prices and costing non-market impacts, but the resulting estimates are problematic in terms of concept, ethical framework, and practicalities' (Stern 2006: 145, 164).

In their calculations, the Stern team focuses on three commensurable values: health, environment and consumption. Commensurable means that losses in one category can be balanced by gains in another category: higher consumption in one world region, in the extreme case, is interchangeable with the loss of livelihoods or even death in another region.[42] Stern and his team explicitly address this problematic: 'Many would argue that it is better to present costs in human lives and environmental quality side-by-side with income and consumption, rather than try to summarise them in monetary terms. That is indeed the approach taken across most of the Review' (Stern 2006: 145).

But in the summary of the study that gained by far the most attention in climate change debates, the consequences of climate change in these three areas are modelled as a loss of GDP.[43] In order to express the consequences in monetary terms, and to make them fit within economic models, the modelling techniques reduce the complexity of the various dimensions of damages and risks. The common denominator is applied to non-market effects of climate change as well. The Stern Review, then, uses a single aggregate metric that includes mortality, environmental damages and consumption changes.

Translating the complexity of climate change and its consequences into monetary terms involves the valuation of ecosystems and livelihoods and im-

---

[42] This approach became highly controversial in the IPCC Second Assessment Report and was rejected by many governments, as the analysis implies that the costs of human life should be different for different countries, depending on their income levels (Barker 2008).

[43] In his *Blueprint* book, Stern complains that the 'aggregate modelling of the Stern Review was presented in just one of the thirteen chapters [...]. In the other twelve chapters the probabilistic approach to outcomes and impacts were set out in some detail, and the costs of different kinds of action were examined. Nevertheless, Chapter 6 attracted a lot of attention because it provided an aggregate statement of costs of action versus costs of inaction' (Stern 2009a: 93).

plies, inter alia, the equation of monetary losses in different world regions: what means less luxury consumption to some may threaten the basic needs of others. The use of global aggregate values leaves disparities in income distribution unconsidered or, at least, cannot serve to problematise them. The only requirement in the Stern Review regarding incomes is that 'future generations should have the right to a standard of living no lower than the current ones' (Stern 2006: 42): future developments that leave the poor poor and make the rich richer fulfill this premise.

To be clear: the approach taken in the Stern Review is among the most progressive economic studies. It rejects high discount rates that would play down future consequences of climate change, and this reflects, at least partially, equity considerations as well: high discount rates are justified through the assumption of growing wealth, which lowers the costs of future spending for mitigation and adaptation compared to present expenditure. This is based in aggregate values of wealth and consumption, and clearly is inadequate for the poorer parts of populations that face economic challenges from adaptation independent of global GDP development. Stern and his team therefore give a higher value to consumption by the poor to address income inequity. 'However, this does nothing to address the violation of comparability i.e. more consumption compensating for death and destruction' (Spash 2007: 710).

The Stern Review thus reduces the complexity of climate change consequences by expressing them in monetary terms. It is this treatment that allows deriving the optimal level of emissions reductions from the expected costs. 'Abatement should take place up to the point where the benefits of further emissions reductions are just balanced by the costs', the Stern Review emphasizes (354). Likewise, uncertainty and the risks of very high temperature increases are expressed in GDP losses of up to 20 per cent, instead of considering the potentially catastrophic consequences from these changes as the limits to economic calculation.

Stern and his team thus dissociate from traditional economics in naming and taking seriously the extent and non-predictability of potential climate change risks, to then factor them back into traditional economic analysis; they 'manage to convert unknown and unknowable futures into events with known probabilities, and miraculously strong uncertainty becomes weak uncertainty. [...] [T]he future no longer threatens irreversible surprise disasters, but rather such "catastrophes" are now known bounded threshold events measured as reduced consumption growth' (Spash 2007: 709).

Opinions vary among economists on whether the Stern approach is still cost-benefit analysis (CBA). Arguing that CBA is useless for climate change due to the uncertainty and risks of catastrophe involved, Barker (2008) acknowledges that Stern and his team shift from traditional CBA to a new inter-disciplinary and multi-disciplinary risk analysis, and argues that this had a huge effect on the treatment of climate change within economics in general: 'It is now acknowledged that the economics of climate change is now more appropriately concerned with uncertainty rather than return [...]' (Barker 2008: 191). He nevertheless dismisses the Stern approach, suggesting that the role of economics should be to help achieve political targets based on scientific evidence and equity considerations at lowest costs, rather than choosing the targets themselves (ibid.).

Likewise, Ackermann (2007) suggests that CBA 'is useful, but it is far from the whole story. It may be as far into the story as the monetary damages can take us. [...] As to the ambition of monetizing the entire range of impacts, and telling the entire story in a single number: give it up' (22). Instead of relying on CBA to compare mitigation and damage costs, he advocates using the Stern perspective to determine the least-cost strategy for preventing unacceptable climate change consequences. Such an approach, in other words, would replace cost-benefit analysis with cost-effectiveness in climate protection (Pearce 2003).

3.3 After Stern: towards a new rationality in climate governance

Notwithstanding all economic considerations, calculations and interests that prevailed in climate change governance before the publication of the Stern Review, and also notwithstanding the various forms of critique that are raised against his approach; it should have become clear throughout this chapter that Stern an his team made an important contribution to framing climate change as an economic challenge.

Dimensions of economic rationality in climate governance

By distinguishing several dimensions of economic rationality in climate change governance, we can highlight more systematically the contribution of the Stern Review to the climate change discourse. The same classification also serves as a heuristic for the analysis of the climate investment discourse.

- Climate change can be considered as an economic problem *first* in the sense of a *problem for the economy* (in the form of the entire national economy or certain sectors). In the Stern Review, this is contained in the calculation of economic consequences of climate change, and is an important dimension of the main message that mitigating climate change is economically rational.

The economic consequences of climate change likewise play a prominent role in adaptation strategies that often focus on the vulnerability of certain sectors.

- A *second* dimension is the *priority given to economic objectives* in formulating climate protection strategies. The Stern Review considers both economic and non-economic risks from climate change and thus does not reduce it to an economic problem solely. Nevertheless, the Review emphasises that response strategies must be compatible with the objectives of maintaining economic growth and reducing (transaction) costs. Crucial is the emphasis on aligning climate protection and economic growth globally, that is, a perspective that allows for economic and social development in developing countries while maintaining the competitiveness of industries and economies in developed countries through low-carbon growth.

- The priority to economic performance is, *third*, coherent but not synonymous with positing that *economic instruments are most effective* in addressing climate change (or, put differently, that these instruments have the least adverse effects). This perspective was fundamental for introducing the flexibility mechanisms within the Kyoto Protocol. The Stern Review and Stern's preceding publications generally support this argument. However, as Stern and others caution, realising the potential of these markets and instruments for achieving climate protection requires enhancing and improving their regulation. Against this background, it will be important to see how the role of carbon markets is addressed in the climate investment discourse.

- A *fourth* perspective that is contained in the Stern Review rather implicitly is the claim that a *fundamental transformation of economies* is needed to successfully combat climate change. However, Stern's following publications make this point much more explicitly. And the need for transformational change to economies raises the question of finance and investment for climate protection as well, and thus contributes to the finance turn in climate governance that will be addressed in the next chapters.

- The *fifth* and last perspective on the economic dimensions of climate change addresses the *economic causes of climate change*. In one sense, this perspective can be found in the Stern Review as well: It describes climate change as the biggest market failure in history and turns to the spotlight on the role of markets in climate governance. However, it then detaches the negative side of markets from their functional core and argues that in their ideal form, markets can and must be a crucial part of responding to climate change. The role of states in creating markets in this ideal form is an important aspect in the climate investment discourse as well.

A new problematisation of the climate challenge

Against this background, we can identify four important ways in which the Stern Review can be expected to effect on climate governance. The *first* crucial issue strengthened by Stern is the focus on the financial challenges related to climate protection. We saw that the study's estimates of climate protection costs are subject to uncertainties and entail normative decisions also. But all objections to the general approach and the choice of parameters notwithstanding, the Stern Review helped establish a global economic perspective that allows to frame consequences of climate change and benefits of climate protection in monetary terms. It 'shifted the debate away from polar bears and unseasonal summers, and reframed it in the cold hard language of the balance sheet' (The Guardian, 30 March 2009).

On this ground, it is possible to describe he fundamental challenge for successful climate protection on a global scale as the appropriate distribution of financial resources between different objectives. Stern identifies four priority areas for low-carbon growth (efficient energy use, combating deforestation, quick spread of existing technologies, and scaled up investment for new, not yet marketable technologies), and advocates to realise climate protection by giving priority to directing finance to these areas.

Second, the focus on the monetary side of climate change permits to chose gross economic output as the central success criterion for climate protection strategies, and to describe climate governance as 'the pro-growth strategy for the longer term' (Stern 2006: ii).[44] Critics lament that the Stern Review does not explain 'why the prospect of human induced climate change is best reflected in GDP at all, why is the problem being framed like this, as "the pro-growth strategy"?' (Spash 2007: 710); others caution that '[i]f we continue to measure "progress" in the perverse way we do, if we continue to tolerate gross inequities in our contemporary social world [...] we will need to get adept at living in a world with a continually warming climate' (Hulme 2006).

While not questioning the desirability of economic growth, Stern nevertheless reframes the issue in a fundamental way. Instead of describing economic growth as the entity to be secured from state interventions on behalf of climate change, Stern emphasises the intrinsic link between the two, and that way rationalises the common pursuit of both goals.

The *third* important issue is the active role that the Stern Review ascribes to governments in reducing emissions and organising the transformation to low-

---

[44] In his 2009 book, Stern explains that in the low-carbon world, 'not only will growth be sustained, it will be cleaner, safer, quieter, and more biodiverse' (10).

carbon economies. Stern and his team call on governments to correct 'the greatest market failure the world has seen' (Stern 2009a: 11), and to put in place the conditions that make markets work for climate protection. While the large majority of climate related activities would have to come from private businesses and investors, 'governments can harness the power of markets to find an effective international response to the challenge' (Stern 2009b: 36).

This way, Stern dissociates from other economists´ more critical perspective on the role of governments for the functioning of markets. This does not mean that Stern advocates limiting the role of markets in combating climate change, all to the contrary: Markets remain at the centre of the solution to climate change. In his 2009 *Blueprint* book, for instance, Stern expresses a clear preference for carbon trading over other instruments and forms of climate regulation. Nevertheless, he claims, a stronger role of the state is necessary to enhance the contribution of carbon markets to achieving climate protection.

Fourth and finally, the Stern Review combines enhanced attention on the urgency of combating climate change with a strong sense of opportunity. Stern emphasises this perspective three years later: 'If we start now and plan carefully, the costs of achieving low-carbon growth will be modest relative to the risks avoided. And we shall discover many new opportunities along the way that are likely to make costs much lower than we now anticipate' (Stern 2009: 38).

This message was and is strongly taken up in climate change debates and is widely used to address concerns with the negative consequences of climate protection efforts. Amartya Sen, Nobel Prize economist, suggests that '[w]hat is particularly striking is the identification of ways and means of sharply minimizing these penalties through acting right now, rather than waiting for our lives to be overrun by rapidly advancing adversities. The world would be foolish to neglect this strong but strictly time-bound practical message'.[45]

The role that the Stern Review plays in closing the gap between the perceived magnitude of the climate threat and the waiting game played by many economists is crucial for explaining the broad appraisal of the report. Stern criticises the shortcomings of many economists to not offer a strategy for ambitious climate change mitigation for economic reasons, and for ignoring the ethical dimension of climate change: 'It is remarkable how many economists make the mistake of shying away from this discussion, engaging instead in fruitless attempts to identify so-called "revealed" ethics and values from the outcomes of real-world markets' (2009: 75).

Stern closes these gaps. By translating risks and uncertainties into economic terms, and by incorporating ethical arguments for climate protection into his

---

[45] http://webarchive.nationalarchives.gov.uk/+/http://www.hm-treasury.gov.uk/d/20061028_Quotes-7.pdf.

calculations, the Stern Review renews the claim that economics can provide adequate analysis and formulate urgently needed answers to the climate challenge. Stern also acknowledges that economic growth led to rising emissions, and that markets failed in giving a price to emissions as a negative externality. His answer is to apply a revised set of economics that recommends a modified form of growth and proper regulation, to overcome market inefficiency and failure. Addressing important weaknesses in the dominant economic perspective on climate change that have been widely felt, it contributes to delegitimising more fundamental forms of critique.

It is in this sense that the Stern Review is an economic problematisation of climate change: It expresses difficulties in the way that the climate domain had been governed up to that point, and translates them into problems that governments can respond to. Instead of radically questioning the current form of economic development due to its contribution to climate change, it suggests minor modifications to this development path that would enable a new phase of growth and prosperity. Through this critique, the Stern Review prepares the ground for new forms of (government) intervention into markets and the economy, and at the same time sets limits to these interventions.

In analysing the climate investment discourse, it will be crucial to describe more in detail the role that is ascribed to governments in making markets work for climate protection, and to see whether this reflects a change to he market-driven climate regime.

4 Making investments an object of government

The previous chapter highlighted how the Stern Review develops and synthesises an economic perspective on climate change and climate protection. Despite the important role that economic analyses, interests and instruments had played in climate change governance long before its publication, the Stern Review contributes to a new awareness for the economic risks and opportunities related to climate change. This growing awareness lifts the issue up the political agenda and initiates a search for new strategies in addressing climate change.

The next two chapters focus on a more specific issue, the climate investment discourse. As highlighted in the introduction to this book, the need and objective to raise private sector investment for mitigation and, to a lesser degree, adaptation activities has gained importance in climate change governance in recent years; an increasing number of agencies deploys instruments to incentivise, or direct investment for climate protection.[46]

The introduction to this book suggested understanding the Stern Review and the climate investment discourse as parts of the same discursive shift. It is thus important to highlight how the growing awareness for the economic risks and opportunities related to climate change is taken up, reflected and transformed in addressing investment for climate protection: The awareness that coping with climate change requires a fundamental transformation of economies, coupled with the increasing visibility of economic opportunities related to rebuilding economies and energy sectors, leads to a pursuit of new strategies for driving investment to the clean technologies and production processes of the future.

Drawing on the terminology developed in the theory section, we can specify the analysis of the climate investment discourse as explaining the emergence, form and significance of new government practices directed at investment for climate protection, and the strategy that informs these practices. This chapter focuses on the first step of this inquiry. It identifies three contexts in which different actors increasingly address investment for climate protection, and thus

---

[46] This is not to claim that climate policy thus far had no effect on the direction, level or types of investment flows, all to the contrary: every form of regulating emissions entails consequences for investment decisions, and investment had been addressed, *inter alia*, through the Clean Development Mechanism (CDM) and renewable energy policies. What is new, however, is the importance of addressing investment for climate protection.

constitute investment as an object of government: the climate finance field; low-carbon economy strategies; and renewable energy policies.

Describing the discourses within these contexts and changes to them allows highlighting the forms of knowledge and political agency related to the government of investment. This will serve, ultimately, to identify the (discursive) structures or rationalities that can explain the particular approach taken for the government of investment; a particular focus in this analysis is on the role that is ascribed to the state in governing investment for climate protection.

4.1 Climate finance: supporting north-south investment flows

The climate finance discourse is fundamental for understanding the approach taken to investment for climate protection. An important change has taken place here in the years since the publication of the Stern Review: The focus shifted from supporting adaptation needs of developing countries to their contribution to global climate protection. This also entails a shift from public transfers for justice and equity reasons to stimulating private investment for climate related activities. As the Advisory Group on climate finance to the UN Secretary General put it: 'International private investment flows are essential for the transition to a low carbon and climate resilient future. These investments can be stimulated through targeted application of concessional and non-concessional public financing' (AGF 2010: 10).

4.1.1 The early days of adaptation and finance

Finance as such is all but a new issue in climate change governance. From early on in the international negotiations, developing countries highlighted their financial needs for preparing for and adapting to the consequences of climate change and asked for support from developed countries. But climate finance was rather subordinated on the political agenda and in public awareness for long, as was the adaptation issue in general. The 1997 Kyoto negotiations that were crucial for the approach taken to climate challenge in the following years clearly focused on mitigation (Schipper and Burton 2009).

The lower importance of adaptation was due to political as much as practical reasons. On the one hand, scientists and activists were long concerned that addressing adaptation might lower the commitment to climate protection and emission reductions. 'For the best part of a decade, discussion of adaptation was regarded by most participants in climate policy-making as tantamount to betrayal' (Pielke, Prins et al. 2007: 597). This led, according to some, to an unnecessary delay in producing knowledge and strategies for adaptation (Dietz 2006).

On the other hand, implementing adaptation is more complex than reducing emissions. Realising mitigation raises manifold political and social challenges as it is often at conflict with powerful vested interests; nevertheless, the possibilities for reducing emissions are in principle well know. In planning and implementing adaptation, to the contrary, problems already arise on the scientific and conceptual level, as the understanding of climate change consequences is still rather limited and prone to many uncertainties, in particular on the regional and local level. Identifying appropriate adaptation measures is further constrained by specific local circumstances and the complex socio-economic dimensions of adaptation.

Nevertheless, the adaptation issue gained ground after the IPCC Assessment Reports in 1996 and 2001 provided evidence that a certain level of climate change is no longer evitable and some of these changes are already happening (Smit and Pilifosova 2001). This was strongly reaffirmed by the Fourth Assessment Report (AR4) in 2007: 'Even if, by some miracle, we could stop emitting greenhouse gases today, we will still experience climate change in the next few decades, making adaptation unavoidable'.[47]

Financing principles: UNFCCC and Kyoto

In relation to this belated attention for adaptation, the UN Framework Convention on Climate Change (UNFCCC) and the Kyoto Protocol entail no detailed provision for providing climate finance; they do give, however, give important advice for negotiators on how to organise a 'global climate finance regime' (Stewart, Kingsbury et al. 2009a: 22).

The climate finance provisions in the Framework Convention and the Kyoto Protocol rest on the distinction between Annex 1 and Non-Annex 1 countries, which, roughly speaking, is a distinction between developed and developing countries. While the industrialised countries as major emitters are seen as mainly responsible for anthropogenic climate change, the (mostly) poorer countries in the south are more vulnerable to its consequences and thus require support.

The Framework Convention reflects these unequal responsibilities and capabilities and translates the principles of equity and fairness into 'common but differentiated responsibilities' (Article 3.1): countries should contribute to cli-

---

[47] Richard Klein, coordinator of climate policy research at the Stockholm Environment Institute and a coordinating lead author of the IPCC, during the presentation of the IPCC Assessment Report 4, cf. http://www.iisd.org/media/press.aspx?id=70.

mate protection according to their historical responsibilities for global greenhouse gas emissions and their present capabilities to reduce emissions.

With respect to climate finance, Article 4.3 of the Convention calls on developed countries to support developing countries in meeting the full incremental costs of mitigation measures. The extent to which developing country Parties will implement mitigation measures depends on the 'effective implementation by developed country Parties of their commitments under the Convention related to financial resources and transfer of technology [...]' (Art 4.7m). But the Convention does not offer further guidance regarding the level of resources and finance to be provided for mitigation, or how to determine this level.

For adaptation, Article 4.4 calls on developed country Parties to assist those developing countries that are particularly vulnerable to climate change in meeting the costs of adapting to adverse effects. Article 4.8 specifies the factors that make countries eligible to such support, which effectively includes all developing countries.[48] Article 4.5 additionally calls for supporting and financing technology transfer to developing countries for both mitigation and adaptation.[49] The Kyoto Protocol does not go far beyond the UNFCCC: in its Articles 10 and 11 it calls for technology cooperation and reaffirms the provision of 'new and additional financial resources' for mitigation and adaptation.

The principles of *incremental* costs and *new and additional* finance are crucial in climate finance and have given rise to intense discussions. Incremental costs means that developed countries do not have to meet the full costs of a mitigation measure in developing countries, say, the installation of a solar power plant, but only the costs additional to business as usual development, for example, energy generation from a coal power plant. It is very difficult, however, to determine these incremental costs in practice.

The principle of new and additional finance causes similar problems: developing countries repeatedly call on developed countries to stick to this principle and not rename existing aid budgets as adaptation finance, which nevertheless has become a common practice. On a more technical level, distinguishing adaptation from other development efforts is difficult, which makes it likewise difficult to distinguish development and adaptation finance.[50]

---

[48] Due to the pressure of mainly the OPEC group of countries, Article 4.8 also calls for the financial support of countries that face negative impacts from the implementation of response measures, in particular those countries whose economies are highly dependent on fossil fuel industries: this provision has led to many conflicts over the years and effectively hampered the operationalisation of adaptation finance.

[49] And Article 4.3 also calls for financial support for national communications and GHG inventories.

[50] Many development agencies, therefore, aim at making development planning (and finance) climate proof.

Adaptation funds: Objectives and governance

The most important step towards realising climate finance under the umbrella of the UNFCCC thus far is the installation of three adaptation funds at the Marrakech COP in 2001 (Schipper 2006). The Adaptation Fund finances concrete adaptation projects and programmes in developing countries that are particularly vulnerable to the adverse effects of climate change. It is funded by a 2% levy on transactions under the Clean Development Mechanism, and could thus become operational only with the entry into force of the Kyoto Protocol in 2005. Three features distinguish the Adaptation Fund from other international finance mechanisms and make it, for many developing countries, a model for a reformed and scaled-up climate finance mechanism.

First, COP 13 in Bali decided that the Adaptation Fund Board be under the authority and guidance of the Conference of the Parties to the Kyoto Protocol (COP/MOP) and fully accountable to it, and managed by the Adaptation Fund Board (AFB) under the umbrella of the UNFCCC.[51] *Second*, the AFB represents country member groups to the UNFCCC in a balanced manner, which results in a developing countries majority within the AFB – a unique constellation within the range of international funding mechanisms.[52] And *third*, Decision 1/CMP.3 gives Parties 'direct access' to the Fund resources: any National Implementing Agency that meets the fiduciary standards of the Adaptation Fund is in principle entitled to receive funds directly.

The other two funds agreed on in Marrakech are operated through the Global Environment Facility (GEF): the Least Developed Country Fund (LDCF) supports these countries in elaborating National Adaptation Programmes of Action; and the Special Climate Change Fund (SCCF) was established to finance projects relating, among others, to adaptation, technology transfer and capacity building.

The Global Environment Facility (GEF) was made the interim operating entity of the financial mechanisms for the UNFCCC in 1992, 'to a certain extent by default, as there appeared to be, in the eyes of many, no alternative' (Gomez-Echeverri and Müller 2009: 2). Though accepting the need for climate change funding in developing countries, the developed countries militated against establishing a new financial mechanism.

---

[51] The COP/MOP decides on the overall policies of the Adaptation Fund in line with further COP decisions that are relevant for adaptation funding (Decision 1/CMP.3para.4, see Parker, Brown et al. 2009: 131).

[52] This is sometimes seen as the reason why making the Adaptation Fund operate has proven difficult, as developed countries are not ready to dedicate considerable amounts of money to a mechanism whose operation they cannot control (personal communication, Annex 1 negotiator, UNFCCC meeting, Bonn, June 2010).

Generally, the decision making process under the UNFCCC is governed by the principle of one country, one vote. This does not hold true to the same extent for the GEF as the executing agency, where the voting rights depend, amongst others, on the size of financial contributions, which gives *de facto* veto power to the largest donors. Altogether, the decision making process for the funds under the UNFCCC represents 'an imperfect and uneasy compromise between donor and developing country power over funding decisions' (Porter, Bird et al. 2008: 47).

What is important, however, is that all adaptation funds were created as a form of compensation for the damages and dangers that developing countries would face from anthropogenic climate change, and thus largely followed justice considerations. This is reflected, inter alia, in the notion of vulnerability and the priority given to Least Developed Countries. The financial flows are meant to meet the additional costs of developing countries from immediate adaptation needs (UNFCCC 2005). Consequently, it was hardly ever questioned that the financial transfers for adaptation would have to come from public sources and should carry no or only weak conditionalities.

The strong general agreement on the function and form that the adaptation funds should have – different perspectives regarding their institutional structure, the access to the funds and the objectives for which the resources could be used notwithstanding – the actual deposit of resources to these funds remained scarce over the years. Developed countries' treasuries proved to be rather reluctant in earmarking taxpayers' money for adaptation policies in developing countries. The financial provision through the adaptation funds has thus been a source of major conflicts within the climate negotiations for many years (Raworth 2007).

4.1.2 The new importance of climate finance

The finance issue moved up the climate change agenda since the Bali COP in 2007, together with a focus shift away from adaptation needs and towards financing climate protection in developing countries; this changing understanding and role opens the climate finance debate for financing strategies beyond mandatory funds.

Various developments and changes contributed to the growing importance of climate finance. One (minor) reason might be the increased importance of adaptation policies, related again to the growing awareness of the vulnerability of developed countries. While the consequences of climate change were seen as a threat to developing countries in the first place when negotiating the Kyoto

Protocol,[53] the awareness of the vulnerability against climate change has risen strongly in northern countries at the beginning of the 21st century, and the countries of the European Union in particular have intensified their adaptation planning for sectors like agriculture, energy production and tourism (European Commission 2007; Jacob 2008). But whereas climate change is an immediate threat to the survival of many people in developing countries, it is rather monetary wealth that is at stake in developed countries (Stern 2009a); nevertheless, vulnerability against climate change is a social phenomenon here as well, and the poorer segments of society are seen as more vulnerable (European Commission 2007).[54]

More than the direct consequences of climate change, many industrialised countries seem to worry about the indirect threats of climate change. Environmental changes in other world regions, from this perspective, can unleash increased migration, political instability and conflicts, or threats to energy supply and transport routes. A number of studies published since 2007 raised awareness of these threats, giving greater concerns to global interdependencies and indirect vulnerabilities (CNA 2007; WBGU 2007). Attempts were made to (re-)frame climate change as a security issue, and – at least in the public debate –security threats are used as an argument for enhanced climate protection efforts.

In the first place, however, it was two other developments that raised the importance of climate finance: on the one hand, the finance issue gained strategic importance for finding agreement between developed and developing countries in the climate change negotiations; on the other hand, a new emphasis on the need for climate protection in developing countries raises the question of how to support these mitigation efforts financially.

The strategic side of providing climate finance

The increasing awareness of the extent of climate change consequences through the IPCC AR4 and others was accompanied by first estimates on the costs of

---

[53] The higher vulnerability of developing countries is due to a number of reasons: the direct climatic consequences are stronger and more variable in developing countries; the dependence of populations on natural resources is often higher in these countries; and the resilience and adaptation capacity are lower than in developed countries (World Bank 2003).

[54] This social dimension of climate change notwithstanding, the direct threats to the industrialised, and especially the European, countries are limited in comparison to developing countries, and are expected to be controllable: according to an early study by the German Federal Environment Agency, more or less complex solutions could reduce vulnerability in almost all cases and regions to a minor degree, and adaptation is expected to bring new economic opportunities as well, in the form of job and market creation (Umweltbundesamt 2005).

adaptation and the financial needs of developing countries, which varied from US\$ 10-40 billion (World Bank 2006) to at least US\$ 50 billion (Raworth 2007) annually. Though rough estimates only, these numbers helped to make the case for financial support for developing countries. For NGOs like Oxfam that were among the first to push the adaptation issue, this was one of the main reasons to put forward such numbers.[55] The need for some form of compensation was widely accepted and supported by the important (international) organisations of development cooperation such as the World Bank as well.

From a developing countries perspective, financial support for adaptation is a justice issue in the first place. The importance as a justice issue also helps to explain the peak of the climate finance topic in the run up to the Copenhagen COP in 2009, reflected not least in the creation of the UN Secretary General's Advisory Group on Climate Finance.

At COP 13 in Bali in 2007, Parties had agreed on a new mandate known as the Bali Action Plan to develop at COP 15 in Copenhagen an agreement that replaces the Kyoto Protocol. The Bali Action Plan once more emphasizes that mitigation in developing countries is tied to financial support and technology transfer, and also renews the call for new and additional finance (UNFCCC 2008). In the negotiations since Bali, developing countries repeatedly insisted on additional financial resources as a fundamental element of any future agreement.

But the Bali Action Plan concretised the finance issue as well by introducing Nationally Appropriate Mitigation Actions (NAMAs): these reflect a compromise between the interest of developing countries to autonomously plan and implement sustainable development policies and projects and receive financial support for these, and the request of developed countries to guarantee the mitigation effect of these policies and projects. NAMAs are thus entitled to technology, financial and capacity building support and are subject to measurement, reporting and verification.

All eyes were then directed to COP 15 in Copenhagen in December 2009: Copenhagen saw the greatest media coverage in the history of climate negotiations and the largest number of heads of governments ever attending an international meeting. Finance was among the most important topics at a COP for the first time, and in the run up to the summit many emphasized that agreement on this issue will be crucial: 'If there is to be a deal in Copenhagen, something will have to give – and it must be the rich countries finance ministries' (Müller 2009a: 2).

Given the tension between the overwhelming expectations that accompanied the Copenhagen conference, and the little scope for far-reaching agreement

---

[55] Interview Tim Gore, OXFAM International.

that many, including negotiators themselves, had warned of, finance became one of the few topics that allowed for progress. Other observers suggest that the finance issue moved to the core of the agenda due to the presence of Presidents and Prime Ministers, as it is one of the few topics within the complex climate architecture that heads of state can negotiate on easily.[56]

In fact, the most substantive step forward made in Copenhagen was the commitment to provide US$30 billion in fast start climate finance for the years 2010-2012, including public transfers and investment from international institutions, plus the *goal* of mobilising US$100 billion dollars a year by 2020 for 'meaningful mitigation actions' from 'a wide variety of sources'; adaptation funding is meant to come from 'new multilateral funding' (UNFCCC 2009).[57]

The finance issue remained crucial after Copenhagen for patching up the relation between developed and developing country parties: one of the foremost challenges after the disappointing outcome of COP 15 was to rebuild trust in the negotiation process (Chasek 2010), and climate finance appeared as one 'crucial sort of trust building block'.[58] On a less positive note, highlighting financical contributions by developed countries is seen as a way to sidestep, at least partly, the equity concerns of developing countries, as no progress could be achieved on emission reductions in Annex I countries as the core of the climate agenda. This interpretation is supported by the emphasis that was given in Copenhagen on financing forest protection, an issue that allows creating a win-win situation for all countries and requires only *financial* commitments form developed countries.

The Bali and Copenhagen COPs did not only raise the attention for climate finance, but shifted the focus away from the pure provision of public funding.[59] The emphasis on innovative means of funding to assist adaptation in developing countries, and on the 'mobilization of public- and private-sector funding and investment, including facilitation of climate-friendly investment choices' (UNFCCC 2008: 5) contribute to a new understanding of climate finance.[60]

---

[56] Interview Ramzi Elias, Project Catalyst.

[57] A significant portion of such funding is expected to flow through the Copenhagen Green Climate Fund, which Parties established as one, though not the only, operating entity of the financial mechanism of the Convention. The Green Fund can support mitigation and adaptation projects, programmes, and policies including REDD-plus, capacity building, and technology development and transfer.

[58] Interview Tim Gore, OXFAM.

[59] It is in particular the introduction of NAMAs in the BAP that provides a framework for making private investment a part of climate finance flows: public funding is meant to support the realisation of mitigation and adaptation measures planned by the host countries, but not necessarily to finance these measures; it can finance readiness activities instead.

[60] For some, the emphasis on private finance was a strategic argument as well, as including private investment in the overall climate finance package made the offer by developed countries look more (Interview Ramzi Elias, Project Catalyst).

Mitigation in developing countries: the cost-effectiveness argument

Besides the strategic relevance of climate finance, it was mainly the increasing need for climate protection in developing countries that helped push climate finance up the political agenda. The rising emissions in some developing countries, in particular in emerging economies like China and India, played an important role for this and were the subject of intense discussions at COP 15 in Copenhagen as well.[61] It is now often claimed that China and the US as the two largest emitters are expected to go ahead with emission cuts and that the new large emitters should accept some form of binding emission reduction commitments.[62] Likewise, it is widely accepted that stabilising emission pathways in other developing countries at least in the long run is crucial for successful global mitigation.

While it is mainly a few major developing countries whose rising emissions raise concern, it is the availability of cheap abatement opportunities that has moved the rest of the developing world to the centre of climate protection strategies: If scarce financial resources are used to finance emission reductions in developing countries, than the overall level ambition of global climate protection could be higher, so the argument goes.

It is important to note that cost-effectiveness as such is all but a new objective in climate politics. The 1992 Framework Convention urges for mitigating climate change 'at the lowest possible cost' (United Nations 1992, Art. 3.3), and the cost-effectiveness principle was institutionalised through the Kyoto Protocol's flexibility mechanisms that allow for financing abatement abroad to meet domestic targets.[63] These mechanisms embrace the win-win rhetoric of ecologi-

---

[61] China and India have become major emitters, at least in terms of national emission budgets. On a per capita basis, industrialised countries remain far ahead and alignment is not likely to happen even within decades, let alone the fact that large shares of developing country emissions are ultimately consumed in developed countries.

[62] The Chinese government was made responsible for the failure of the Copenhagen summit by rejecting a proposal from a group of industrialised countries in the last hours of the summit. These debates reveal the different understandings of responsibility for climate protection: Annex I governments that should act first on climate protection according to UNFCCC principles had little to offer to developing countries in Copenhagen regarding emission reductions or financial support, and it was the same countries that went ahead with backroom diplomacy after many developing countries refused to agree to a deal without such commitments. Much of the public opinion, however, blamed China for not agreeing to the proposed deal and for the 'hijack' of the Copenhagen Climate Change deal (Ed Miliband, UK Climate Secretary, cf. http://www.guardian.co.uk/environment/2009/dec/20/ed-miliband-china-copenhagen-summit?intcmp=239).

[63] In the Kyoto negotiations, the cost-effectiveness argument was mainly pushed by a US led coalition of industrialised countries that were widely reluctant to take binding emission reduction commitments (the members of the JUSCANZ group were Japan, the US, Canada, Australia, and New

cal modernisation, they are supposed to achieve the double objective of sustainable development investments in developing countries and low cost mitigation for industrialised countries (Bäckstrand and Lövbrand 2006).

However, the flexibility mechanisms were not meant to be at the heart of the Kyoto architecture initially. The Protocol entails a clear commitment to focus on emissions reductions in developed countries, and the flexibility mechanisms should supplement this domestic action (United Nations 1998, Art. 17).

Against this background, the climate finance debate makes explicit a development that has taken place since the adoption of the Kyoto Protocol: north-south finance for climate protection has moved from the margins to the centre of the political agenda, and reducing emissions in developing countries has become an objective in itself rather than complementary to mitigation in developed countries.

This shift to global cost-effectiveness as the primary principle is embedded within a rationalization of the cost-effectiveness argument. One important tool of rationalization has been McKinsey's Global Greenhouse Gas Abatement Cost Curve (GGGACC) that became very popular in climate change discourse and governance in recent years. [64]

The GGGACC takes a global economic perspective on climate change mitigation that is comparable to the Stern Review. Different from Stern, however, the Mc Kinsey study does not compare mitigation costs to climate change damages; instead, it calculates and makes visible the comparative costs of emission reductions in particular sectors or world regions.

While the cost curve thus makes visible in a very accessible way the potential contributions of different world regions to global emission reductions, it is important understandings and assumptions that are not visible in these graphs. The elaboration of the cost curve thus deserves some more attention.

Departing from the insight that mitigation efforts 'close to the full potential' are necessary to remain within 2 degrees of global warming (Mc Kinsey 2009: 6), the cost curve compares the potential of different sectors and world regions to contribute to this objective, regarding their incremental costs or cost-effectiveness. This approach escapes parts of the criticism to the Stern Review, as it does not compare costs to benefits to determine the optimal level of emission reductions, but starts from 2 degrees as the objective defined in the political

---

Zealand), and it was in particular the US who made its endorsement of the Protocol dependent on the introduction of flexibility mechanisms.

[64] The study was prepared by the global consultancy McKinsey in a common initiative with Think Tanks, companies, and NGOs, including the Carbon Trust, Climate Works, Shell, Vattenfall, Volvo, and the WWF. It was financed by Project Catalyst, European partner of the US based foundation Climate Works.

process.[65] Nevertheless, this form of calculation has strong performative effects on the climate finance debate.

The authors of the McKinsey study identify four potential ways of emission reductions: energy efficiency, low-carbon energy supply, terrestrial carbon (forestry and agriculture) and behaviour changes. Taken together, the first three categories allow for emissions reductions of 80Gt CO_2 in 2030. Behaviour changes can deliver another 4Gt. These results rest on a set of assumptions and normative considerations. As the cost curve aims at the most efficient reduction opportunities, it makes no difference where these reductions take place; no requirement exists for a global adjustment of per capita emissions in the long run; and the authors only consider mitigation options 'that do not affect the lifestyles of individuals' (McKinsey 2009: 36).

This last premise is the main reason for the limited abatement potential through behaviour changes in industrialised countries, as '[c]hanging behaviour is difficult [...] and there is a high degree of uncertainty in these estimates' (McKinsey 2009: 29). The authors identify a much larger abatement potential in the forestry and agriculture sector in developing countries. Steering these processes is said to be challenging as it requires 'educating and mobilizing billions of farmers around the world to change their daily practices' (ibid. 34), but the study does not question the feasibility and appropriateness of these changes in general.

Reducing the complexity of abatement opportunities to four categories, the authors can conclude that two thirds of global emission reductions should take place in developing countries due to cost-effectiveness reasons. Two main reasons are given for these comparative cost advantages. Developing countries have the opportunity to 'leapfrog': they can leave out environmental harmful stages of development, and jump directly into more sustainable levels. This level switching is expected to be less costly than rebuilding the industrial sectors in the north. The second extensive and low price abatement opportunity can be found in the forestry sector, through aforestation and avoided deforestation.[66]

Successful implementation of the cost curve is compatible with quite diverse *per capita* emissions in the long run. For the year 2030, the study calculates *per capita* emissions of 7.7 tons CO_2 in industrialised countries; popula-

---

[65] As a political rather than a scientific threshold, the 2 degree goal became popular in the run up to the Copenhagen summit in 2009, and though rejected by many academics and civil society organisations and few Parties to the UNFCCC, it gained almost official status through the Copenhagen Accord and was reaffirmed by governments at COP 16 in Cancún in 2010.

[66] The estimates of 'feasibility and costs' of abatement in the forestry sector are 'subject to significant uncertainty', as 'capturing these opportunities would be highly challenging', as 'More than 90 percent of them are located in the developing world, they are tightly linked to the overall social and economic situation in the concerned regions, and addressing the opportunities at this scale has not before been attempted' (10).

tions in China and India would emit 3.7 tons; in developing countries with a significant share of forestry this value is expected to be 1.9 tons.[67] The authors see no contradiction to equity perspectives, because reduction efforts in the south are expected to be financed by developed countries. Living on less carbon, then, is finally a question of financial support.

The results of the McKinsey study are strongly taken up in the climate finance debate (Spence 2009); many other studies and policy papers draw on the cost-effectiveness argument in particular. The argument for cost-effective abatement in the forestry sector is broadly supported (Stern 2006; Eliasch 2009) and played an important role to move forest protection to the centre of the climate politics agenda (see chapter 5). What is crucial to see is how an economic argument – derived from the comparison of incremental costs – is gradually transformed into an absolute argument in many cases.

While McKinsey emphasizes that two thirds of all abatement opportunities up to a cost of US$60 per ton CO2 are available in developing countries, the *Little Climate Finance Book*, for instance, suggests that of the 17 billion tonnes of emissions reductions required in 2020, '70 % is achievable in developing countries'. Only in a footnote do the authors explain that this calculation rests on the assumption that 'developed countries achieve their full abatement potential up to a marginal cost of US$90/tCO2' (Parker, Brown et al. 2009: 18).

What follows quite immediately from the comparative cost advantages is that a higher share of the financial resources used globally for climate protection is needed in developing countries. A background paper for the German Development Bank (KfW) suggests that more than half of the net incremental investments globally will have to go to developing countries, including to the forestry sector (Scholz and Schmidt 2009).[68] Many other studies support this argument with similar figures (Doornbosch and Knight 2008; Project Catalyst 2009b).

The cost-effectiveness argument thus challenges the fundamental principle of the UNFCCC and the Kyoto Protocol. While this happens implicitly in most cases, Pendleton and Retallack (2009) suggest that 'in a world of limited finance, reductions can arguably be undertaken wherever they can be made for the lowest cost', and offer an alternative perspective on burden sharing: 'Since emissions reductions in developed countries are insufficient to solve the climate problem, [...] the principles of responsibility and capability might more productively be applied to the *financing* of global reductions' (4). In this perspective, it is possible to separate efficiency from equity concerns, and equity concerns are met as

---

[67] The particular low value for these regions results from the fact that avoided deforestation is equalled with negative emissions, resulting in small overall emissions budgets.

[68] The German Kreditanstalt für Wiederaufbau (KfW) is one of the largest climate finance lenders among the bilateral financial institutions.

long as developed countries *finance* the majority of abatement measures (Ackermann 2009).

The McKinsey study and its discussion is therefore an outstanding example for the *performativity of economics* (MacKenzie and Millo 2003): what all the arguments share is that they make an economic argument, often based on rather rough calculations, to an absolute limit for emission reductions in developed countries – a perspective shared by the authors of the McKinsey study as well.[69] But whether 'plentiful cost-effective emission reductions' are available in industrialised countries depends on the particular understanding of cost-effectiveness, and how much importance is given to it relative to other criteria.

4.1.3 Climate finance as investment

Leaving behind the rationality of the Kyoto Protocol this way, however, is not generally accepted, as the Climate Change Conferences in Bali, Copenhagen and Cancún once more demonstrated. Both Non-Annex I countries and many NGOs called on developed countries to move first on emission reductions here, and framed the issue in terms of responsibility rather than simply a question of efficiency in mitigating climate change.

Nevertheless, the emphasis on enhancing mitigation efforts in developing countries, and thus on the need for higher levels of climate finance, plays a central role in making investment an object of government. Comparisons of the required levels of finance with what is available from existing funds and financial mechanisms highlight the need to scale-up climate finance levels substantially – and open the debate towards alternative sources of funding. It is within the framing of public and private contributions to climate finance that a particular role is ascribed to governments in supporting private investment for climate related activities

The need to scale-up climate finance

We already saw that the World Bank and Oxfam were among the first to call for scaling-up adaptation finance and to put forward concrete numbers on the financial needs for adaptation. These numbers were important in raising the awareness of adaptation finance though, as Oxfam suggests, due to the many uncertainties involved in implementing adaptation, 'no one knows how much it will cost for

---

[69] Interview Ramzi Elias, Project Catalyst.

developing countries to adapt to climate change' (Raworth 2007: 17). Consequently, cost estimates vary widely from US$10-40 billion (World Bank 2006) or at least US$50 billion annually (Raworth 2007) in early studies, to the more recent calculations of US$86 billion annually in 2015 (UNDP 2007), US$49-171 billion UNFCCC (UNFCCC 2007), or US$75-100 billion by 2020 (World Bank 2009c; Gore 2010).

Adaptation finance needs depend not solely on climatic changes, but also on the vulnerability of a particular group, and thus on economic, institutional and individual capacities and changes to it. This 'double exposure' (O'Brien and Leichenko 2000) means that resilience building and adaptation are heavily intertwined with socio-economic development, and depend on political decisions in a variety of sectors. Vice versa, policies and measures that target the vulnerability and resilience capacity of a community or population, or aim at adapting societies or sectors to climate change, often have social and economic side benefits. This interdependence of climate vulnerability and socio-economic development makes it difficult to forecast the costs of adaptation, in particular in the long run (Ackermann 2009).

Estimating mitigation costs seems to be simpler at first sight, in particular when mitigation is understood as technology shift. Comparing available estimates is difficult, however, as most studies calculate these costs as a fraction of GDP; the results depend on a number of additional variables like economic growth, population size and technology development therefore, and are prone to uncertainty in this regard (see chapter 3.2). Additionally, these studies set different objectives regarding the concentration of greenhouse gases in the atmosphere, and thus do not only entail a fundamental normative decision on the desired emission path, but operate with different basic variables as well.

The most prominent figures in the debate come from the Stern Review that calculates the annual costs of limiting global warming to 2 degrees (that is, a stabilization of GHG concentrations at 500-550 ppm CO2e in 2050) to be around 1% of GDP (although the estimates of the report range from − 1% (net gains) to 3.5% of GDP); and the 2007 IPCC Assessment Report that estimates the costs of stabilizing GHG between 445-535 ppm CO2e at 3% of global GDP. Using absolute numbers, the UNFCCC secretariat estimates that it will cost US$380 billion in 2030 to return emissions to 2007 level.[70]

While factors such as economic and population growth and the availability of new technologies already limit the accuracy of these calculations, some studies also consider the *policy costs* of realising abatement (Pendleton and Retallack

---

[70] The authors of the study emphasize that this sum is huge compared to the current investments into climate protection, but is only a fraction of 1.1 to 1.7 of total investments and financial flows in 2030 – an important rhetoric shift that is taken up by many actors thereafter (UNFCCC 2007).

2009). This suggests that the costs of mitigation exceed the price difference between alternative technologies, as political interventions are often needed to provide the frameworks for technology deployment or to motivate switching to a new technology;[71] also, realising cost-effective mitigations would require strong coordination across borders that is not always possible. The difficulties in balancing costs this way leads to a *political* cost curve that differs substantially from an *economic* cost curve (Stewart, Kingsbury et al. 2009a).

All those difficulties notwithstanding, projections of global mitigation costs increasingly converge around between US$100-200 billion for developing countries and US$200-400 billion globally by 2020-2030 (Pendleton and Retallack 2009).

More important than the concrete numbers, however, it the broad agreement that financial support for both mitigation and adaptation must be strongly increased, as the finance needs by far exceed what is available from existing climate finance mechanisms. This triggered two parallel developments: proposals for reforming the established climate finance architecture under the UNFCCC also consider how to achieve the required levels of finance, including through new funding sources; more in general, a broader perspective has evolved in recent years on the potential contribution of the private sector to funding climate related activities, and on way to enable and enhance this contribution.

A reformed climate finance architecture

After 2007, governments and international organisations created several new climate funding mechanisms, including from the World Bank, the GEF, and various national ministries. This has led to a fragmented, decentralised climate finance landscape, and to discussions on how to reform or renew the institutional arrangements for a more coherent approach (Porter, Bird et al. 2008). The focus is on the political dimension of this debate here, which is captured in the controversies on the use of *new or existing institutions,* and on whether or not to place the climate finance architecture *under the authority of the COP.*[72]

---

[71] The non-realisation of energy efficiency projects despite negative costs is often described as principal-agent problematic in that sense, for example in the housing sector: landlords refuse to pay for higher energy efficiency of their buildings as it is first of all the tenants that benefit from decreasing energy bills, while the tenants are unwilling to invest in the energy standards of a flat or house that they do not own. Solutions to this problem would result in real costs that deviate substantially from the economic costs (Brinkman 2009).

[72] A rather technical or problem-driven debate on the reform of the climate finance architecture addresses issues like competing authorities, duplication of funding activities and difficulties for developing countries that face an array of uncoordinated founding sources.

The question of COP authority was central when establishing the Adaptation Fund (AF) and its Board (AFB). Developing countries insisted on putting the AFB not only under the guidance of the UNFCCC, but also under its authority.[73] This is seen by many as a reaction to the dissatisfying performance of the Global Environment Facility (GEF) as an operating entity of the UNFCCC financial mechanisms, especially with respect to the non-observance of COP guidance through the GEF (Möhner and Klein 2007).

The same discussion recurred on the establishment of a new and integrative climate finance mechanism for the UNFCCC. While developed countries prefer a governance structure similar to the relationship between GEF and COP, that is, a relation of guidance and accountability, the G77+China insisted that the new finance mechanism and its Executive Board should be under the *authority* of the COP (Müller 2009c).

One crucial dimension in the debate on COP authority is which finance flows are counted towards Annex I commitments, as the UNFCCC references bilateral, regional and other multilateral channels for the provision of climate finance, but is ambiguous to a certain degree on the legal status of these resources. In their proposal for an *Enhanced Financial Mechanism,* the G77+China and the African Group of countries therefore called on developed (or Annex I) countries to spend no less than 0.5% of their GDP to this mechanism to reflect their commitments under the UNFCCC (FCCC/AWGLCA/2008/MISC.2/ADD.1); any other funding would not fulfil these commitments (UN Economic Commission for Africa 2009).

A common proposal by the Mexican and Norwegian governments for a *World Climate Change Fund* that was largely used as a model for the Copenhagen Green Fund reflects some of these concerns (FCCC/AWGLCA/2008/MISC.2): All countries would contribute to the fund financially in order to overcome the traditional donor-recipient structure and give greater responsibility, and therefore voice, to developing countries. To respect the principle of *common but differentiated responsibilities* and the *capabilities* of countries, three indicators (GHG emissions, population numbers, GDP) determine the respective contributions and shares of individual countries, and make sure that developing countries are net recipients.[74]

---

[73] While, according to Müller, there is no official definition of 'under the authority of the COP' within the UNFCCC context, the repeated reference to the relation of AFB and COP suggests that an implicit definition can be found in the respective Decision CMP.3/1 on the Adaptation Fund (Müller 2009c).

[74] This proposal is similar to the Greenhouse Development Rights (GDR) Framework with its 'Responsibility Capacity Index' (RCI), developed by organisations from academia, business and civil society. The GDR pursues a fair formula to distribute the financial burden for mitigation and adaptation to individual countries by taking into consideration their economic capacity and responsibility

The debate on whether to use *new or existing institutions* for climate finance likewise is concerned with the type of and control over climate finance flows. Developing countries claiming a new finance institution usually advocate a strong link to the UNFCCC in order to have a strong voice in finance decisions. Many developed countries, to the contrary, prefer using existing institutions like the World Bank and the Multilateral Development Banks (MDBs) that give more influence to donor countries. This is seen as one important reason why the G8 countries decided to give the World Bank a strong mandate in climate finance through the creation of the Climate Investment Funds (CIFs).

Ultimately, the demands of both developing and developed countries were fulfilled to a certain degree through the creation of the Copenhagen Green Fund in 2009: Developed countries significantly scaled-up their pledges for climate finance flows through the UNFCCC, while the development of a NAMAs registry allows for measurement, reporting and verification (MRV) of mitigation measures in developing countries that are supported through finance, technology and capacity building. AT COP 16 in Cancún one year later, governments agreed that the Green Climate Fund will be accountable to and operate under the guidance of the COP rather than being under its direct authority (Climate Focus 2011).

Public and private money

The creation of the Copenhagen Green Fund is an expression of the new importance of climate finance, but it also entails a fundamental change to the understanding of climate finance and its provision. As we have seen, many developing countries insist on counting only financial flows from Annex I country budgets towards climate finance commitments. The Copenhagen Accord and the agreement on a Green Fund at COP 16 in Cancún, however, highlight that climate finance will have to come from different sources, including the public and private sectors. The proposal for a Green Fund, like many others, lists potential sources of climate finance, including the taxation of emissions trading, levies on international air travel, and auctioning emission allowances at the international level.[75]

---

for emissions. According to this formula, the United States carries the heaviest burden (33.1% of global financial needs in 2010), due to its high emission levels, followed by the European Union (25.7%). The payments of Annex 1 countries would decline in the years thereafter (from 76.9% in 2010 to 60.9% in 2030), and middle-income countries would have to step in.

[75] The *Little Climate Finance Book* provides an overview of these proposals (Parker, Brown et al. 2009).

Proposals for climate finance contributions other than from public budgets in developed countries also raised the question for private sector climate financing more in general – and gradually led to a whole new perspective on climate finance. The discussion on public and private climate finance is the context in which private investment for climate protection is constituted as an object of government.

Once again, the starting point for these considerations are the estimates of climate finance needs. Many studies and strategy papers, in that sense, agree that the required level of climate finance cannot be achieved from public budgets, and that the majority of climate finance flows will have to come from the private sources (UNFCCC 2007; Doornbosch and Knight 2008; Pendleton and Retallack 2009). Other studies also argue that private investment for climate related activities is needed beyond carbon finance, as carbon markets likewise can only deliver a fraction of what is needed (Clifton 2009; Project Catalyst 2009a).

These claims are specified in different ways. A very common approach is to distinguish climate related activities that need public finance support from those that attract the interest of private investors. On the most general level, this is a distinction between mitigation and adaptation. The Stern Review suggests that 'in many cases, market forces are unlikely to lead to efficient adaptation' (Stern 2006: 411), as markets require a level of certainty that is in many cases not available in addressing the consequences of climate change. And adaptation is often about long-term developments or securing public goods, including capacity building, infrastructure development or social security programmes, and thus is unlikely to be pursued by the private sector and market forces (Fankhauser 2006).

Additionally, it is the poor who are most severely affected by climate change, and these often face financial constraints that prevent commercial adaptation solutions. Climate change could therefore 'exacerbate existing inequalities by limiting the ability of poor people to afford insurance cover or to pay for defensive actions' (Stern 2006: 413). In that sense, public finance for adaptation is required to target specific groups and populations: 'It is the world's poorest and most vulnerable people – on the front line of the climate crisis – that adaptation finance must reach. [...] The interventions needed [...] will not attract investment from the private sector, since they do not generate internal returns' (Gore 2010: 3).

In his *Global Deal* book, Stern (2009: 179) emphasizes the need for public climate finance for adaptation and forestry measures and rejects claims of budget constraints in developed countries, arguing that these spending are a rational investment: 'An expenditure on climate security as a global threat, of one-tenth of that on security from external threats from other nations does not look excessive given the danger'.[76]

The picture is different in the case of financing mitigation activities in developing countries. Most strategy papers suggest a mix of public and private money here, with the private sectors delivering the largest shares. One major distinction is between renewable energies and energy efficiency on the one side and forestry on the other side. Whereas the private sector needs to provide the majority of investments for low-carbon energy technologies as well (Doornbosch and Knight 2008; Scholz and Schmidt 2009), the case is different for the forestry and agriculture sector where more public support is needed.

Altogether, the World Bank estimates that over 80% of the investment flows for both mitigation and adaptation are available from the private sector (World Bank 2008a). McKinsey, like many others, calculates the total upfront investment needed for abatement (EUR 530 billion annually in 2020 and EUR810 billion in 2030) and suggests that this is 'within the long-term capacity of global financial markets' (Mc Kinsey 2009: 6).

However, discussions on the private and public shares of climate finance raise important equity concens as well and have been a source of tension in the UNFCCC negotiations and beyond in that sense. Developing countries emphasize that providing climate finance is not just a question of goal achievement and the effective support of mitigation and adaptation, but a justice issue and an obligation under the UNFCCC; at least some of these countries have thus repeatedly rejected private climate finance as a distraction from this justice dimension.[77]

Enabling investment for climate protection

Nevertheless, the issue of private climate finance gained track in particular in the run-up to the COP 15 in Copenhagen in 2009. Several organisations and networks pushed hard to place the topic on the political agenda, and met great interest from national delegations. The sponsors of the Mc Kinsey cost curve, for instance, had published two studies on the climate finance issue that consider the role of private investment, and also organised several events on the issue in Copenhagen; and Nicholas Stern, lead author of the Stern Review, had organised a high-level public-private dialogue in the eve of the Copenhagen summit that

---

[76] He suggests, however, that earmarking financial flows from CO2 taxes or auctions of emission certificates could increase the acceptability of a global financial deal.
[77] Interview Nick Robins, HSBC, and Tim Gore, Oxfam.

considered ways to engage the private sector for financing climate related activities.

Though not an official part of the intergovernmental negotiations, there were various channels through which these perspectives found their way into the Copenhagen agenda: Nicholas Stern and several of his colleagues were advisors to governments in Copenhagen, and some of those who took part in the public-private dialogue organised by Stern also became members to the High-Level Advisory Group on Climate Finance that the UN Secretary General set up shortly after Copenhagen; likwise, the Mc Kinsey cost curve had become a very popular tool for planning national climate strategies, and the experts of Mc Kinsey and Project Catalyst ran consultancy meetings with a great number of national delegations.

Many other organisation also ran side events in Copenhagen that addressed the finance issue, and in particular the role of the private sector. Titles such as 'Financing the battle - scaling up private sector investment through public mechanisms', 'Breaking Barriers to Private Sector Investment', or 'What the private sector needs from Copenhagen to enable finance to flow to low-carbon technology' say a lot about the climate change discourse at that time: Achieving climate protection, apparently, had turned into a financial challenge, and the private sector had taken a new an central role in these considerations.

One important driving force of these discussions were the World Bank and the Multilateral Develoment Banks (MDBs). In 2007, the G8 had given a mandate to the World Bank for the creation of several Climate Investment Funds (CIFs), and had pledged high levels of finance to these funds.[78] In their decision the G8 endorsed the Bank's experience and capability to handle large investment projects.[79]

In line with other World Bank lending facilities, the CIFs aim at a broader approach to financing climate related activities, an in particular at the participation of the private sector. The World Bank, together with its MDB partners, thus

---

[78] According to a sunset clause in the project document, the CIFs are meant to be an interim financial mechanism only that ceases operating once a new financial mechanism under the UNFCCC is in place (World Bank 2008b).

[79] The G8 also endorsed the capacity of the World Bank to quickly create the structures that are needed to enhance climate finance flows. Many developing countries, together with civil society organisations, reject a stronger role of the Bank exactly for this experience, due to the environmental performance of past and current World Bank lending and its governance structures, in particular the voice and vote given to developing countries in the decision making processes (Bretton Woods Project 2008; Third World Network 2008). However, the governance structure of the CIFs gives more influence to developing countries than other World Bank lending facilities, which is seen by some civil society observers as a reaction to the critique of the lending practices of both the GEF and the World Bank (Interview Susanne Breitkopf, Greenpeace UK).

encouraged debates on new public-private investment initiatives for climate protection.

It is crucial to see how the issue of private climate finance is framed and addressed in these discussions and strategy papers. Private investment for climate protection is understood as part of overall climate finance flows; rather than to ask for the particularities of private spending for climate related activities, the focus is on how to enhance the overall level of climate finance through enhancing private sector contributions.

It is within this perspective on public and private contributions to climate finance that a particular role is ascribed to governments for addressing investment for climate protection. Part of this is the way in which public money is meant to be spent: Governments should catalyse investment for mitigation (and, to a lesser degree, adaptation) by using public money; public finance, in that sense, is used as a *lever* or *leverage instrument* for encouraging private investment (UNEP Finance Initiative 2009).

While both public and private money contribute to overall climate finance levels (and are often equated with respect to this purpose), the *golden rule of public funding* requires that 'governments should only support those investments that are economically efficient but not financially viable' (Doornbosch and Knight 2008: 24). This means that public spending should realise projects that are on the left side of the cost curve but for a variety of reasons do not attract the interest of investors, 'those projects which the market forces will not deem profitable' (Schalatek 2009: 22).

More in general, then, an investment perspective on climate related activities points to the particular challenges that investors face, and to the difference between the costs and the investment needs of climate related activities. In a certain way, this reflects the difference between the Stern Review and the Mc Kinsey cost curve as well. While Stern deploys a global cost perspective on climate change and its mitigation to make the general argument that climate protection is economically rational, the McKinsey study highlights that the investment needs for climate protection are different: Investors are not interested solely in the economics of a project, but factor additional criteria into their decision, and might thus choose projects with lower capital intensity rather than with the lowest costs (Mc Kinsey 2009).

Comparing the costs of climate protection measures is useless to a certain degree to inform investment decisions then, as 'investors don't live in models'.[80] Instead of studying the economics of climate change, the role of governments would be to focus on the practical challenges that investors face and support

---

[80] Tom Burke, Founding director of E3G, at the conference *The great transformation – Greening the economy*, at Heinrich Böll Foundation Berlin, Friday, May 28th 2010 (own documentation).

them in realising climate related projects, in particular in developing countries. Financial incentives are part of this support: In the finance terminology, the role of public finance is to crowd in private investments by creating supportive investment conditions rather than to crowd out or compete with these investments (World Bank 2008a).

This investment perspective on climate related activities is fundamental for the approach to the government of investment for climate protection that has developed in recent years, and will be taken up in the next chapter again.

For now it is important to see how the climate finance discourse contributes to making investment an object of government. The need to address investment for climate protection develops gradually within the perspective of financing climate related activities in developing countries. The calculation of the costs and investment needs for the required mitigation activities in developing countries leads to the conclusion that private investment is crucial for climate protection in developing countries, and governments have to play an important role for enabling these investment flows.

Before having a closer look at the World Bank CIFs and other instruments for governing investment for climate protection, the two following sections describe how investment is addressed within low-carbon growth and development strategies in both developing and developed countries (4.2); and how investors take up the challenge of financing climate related activities, and contribute to spelling out an investment perspective this way (4.3).

4.2 Low-carbon strategies: Investing in economic future

About the same time that finance became a more important issue in climate change governance, governments increasingly started developing strategies for a transformation to low-carbon economies, as a way of aligning economic growth with climate protection. The low-carbon transformation requires the development and deployment of new infrastructures, technologies and production processes, and thus is, according to the World Economic Forum, not last a question of finance: 'To ensure our future prosperity, we need a high-growth and low-carbon economy. To that end, a set of practical policies and incentives is urgently required to help remove the obstacles to more low-carbon finance'.[81]

The low-carbon economy discourse gathered momentum through proposals for a Green New Deal that were a reaction the global economic and financial

---

[81] World Economic Forum Task Force on Low-Carbon Prosperity, CEO Climate Policy Recommendations: http://www.weforum.org/issues/climate-change.

crisis in 2007, and the financial stimulus programmes that most major economies put forward in 2008 and 2009.

Consequently, the low-carbon economy discourse inherits a strong sense for economic opportunity from Green New Deal proposals (Luke 2009). Whereas climate protection objectives are often the starting point for low-carbon strategies, it is increasingly the economic benefits expected from this transformation that drive government activities. This becomes apparent in particular in countries like China that are less explicit on climate protection objectives but nevertheless are among the frontrunners in putting in place regulation to drive investment to renewable energies and clean technologies.

This section on the low-carbon discourse starts with the *Green New Deal* (GND) debate that highlights the opportunities related to creating green economies but largely focuses on greening financial stimulus packages (4.2.1). *Low-carbon economy* strategies, to the contrary, shift the focus from green short-term recovery programmes to a more fundamental transformation of economies (4.2.2); having a closer look at some country cases highlights the common rationality in addressing investment for the low-carbon transformation, but also important differences in the strategies that these countries pursue (4.2.3).

4.2.1 A Green New Deal

Low-carbon strategies and GND proposals share the concern for aligning environmental or climate protection with economic objectives. Nevertheless, they take different starting points: whereas low-carbon strategies formulate concrete policy proposals for achieving climate protection objectives and economic growth simultaneously, and are thus largely expert-driven, the idea of a GND caused wider public debates as a reaction to the financial crisis and concerns for a global recession after 2007.

As the name Green New Deal suggests, the idea originated in the US. It is reminiscent of the New Deal that US President Theodore Roosevelt implemented after the Great Depression in the 1930s to revitalize a staggering US economy through public investment and a series of fiscal reforms and economic programmes.[82] The merits of this state-led economic recovery programme remain disputed, the New Deal is an 'ambiguous symbol [in a] society that rarely professes to embrace state interventionism, bureaucratic aggrandizement or economic planning' (Luke 2009: 14).

---

[82] Germany and Italy under the Hitler and Mussolini governments deployed similar state-corporate policies that sometimes are coined New Deal as well (Schivelbusch 2007).

The basic idea of a Green New Deal is to solve environmental problems through redirecting money to green growth orchestrated by government interventions. This idea is a recurring theme in a variety of programmes and theories in the US and beyond: the sustainable development concept that answers to the limits to growth and seeks to harmonise economic growth with social and environmental objectives grants a stronger role to government and policy in setting standards for market activities; a more active role of the state is also an important issue in proposals for green statism (Eckersley 2004); and a variety of concepts such as *cradle to cradle* and *eco-efficiency* propose forms of natural capitalism that replace the current destructive version of capitalism.

A more muscular green

The Green New Deal concept itself is of more recent origin; it was first put forward about ten years before the 2007 economic and financial crisis hit countries worldwide. In his book *Earth Odyssey*, Mark Hertsgaard suggests to revive New Deal policies in form of a *Global Green Deal* that relies on market mechanisms to the extent possible, but cautions that governments must establish the 'rules of the road that compel markets to respect rather than harm the environment' (Hertsgaard 1998: 330). A Global Green Deal could contain tax reforms, subsidies and public investment to push the development of solar power and other renewable energy industries.

Such a transformation, however, requires that governments 'steer in a fundamentally different direction than at present', primarily by redirecting military spending to environmental protection; this, however, 'will upset those who profit from the status quo' (ibid. 333). Additionally, the Global Green Deal should shift wealth from rich to poor, both internationally and within countries.

Now focusing on the domestic situation in the US, Heertsgaard suggests in a 2001 article 'a government-led, market-based plan that will solve the nation's energy problems while also yielding economic returns and addressing the most urgent environmental hazard of our time, global climate change' (Hertsgaard 2001). The centrepiece of this *New Green Deal* proposal is to redirect subsidies away from fossil and nuclear fuels and towards renewable energies, as government spending would create more jobs this way.

The topic was taken up in various major US and UK newspapers and journals. One of the loudest – and most illustrative – advocates of a Green New Deal is the *New York Times* columnist Thomas Friedman. An extensive feature for the *New York Times Magazine* reveals why the green economy perspective attracts the interest of very different constituencies – and why it has nothing to do with a

20th century type of environmentalism that is critical about economic growth and advocates lifestyle changes: 'One thing that always struck me about the term "green" was the degree to which, for so many years, it was defined by its opponents — by the people who wanted to disparage it. And they defined it as "liberal," "tree-hugging," "sissy," "girlie-man," "unpatriotic," "vaguely French"' (Friedman 2007).

Friedman thus suggests reframing the green issue as geostrategic, geoeconomic, capitalist and patriotic. 'I want to do that because I think that living, working, designing, manufacturing and projecting America in a green way can be the basis of a new unifying political movement for the 21st century. A redefined, broader and more muscular green ideology is not meant to trump the traditional Republican and Democratic agendas but rather to bridge them when it comes to addressing the three major issues facing every American today: jobs, temperature and terrorism' (ibid.).

Beyond the terrorism bit of the argument (that can be summed up as: lowering the spendings for oil imports will lower the financial support for terrorist activities as well, or at least for the countries where these activities grow), the call for a more *muscular green ideology* aims at addressing simultaneously the climate and economic agendas of the future. Investing in a cleaner economy allows not only to tackle climate change, but to strengthen economic growth and prosperity through a new cycle of innovation as well – and, ultimately, to maintain the American Way of Life in the 21st century: 'I am not proposing that we radically alter our lifestyles. We are who we are – including a car culture. But if we want to continue to be who we are, enjoy the benefits and be able to pass them on to our children, we do need to fuel our future in a cleaner, greener way' (ibid.).

Beyond the domestic perspective, Friedman argues that low-emission innovation in the US – or western industrialised countries in general – is directly linked to the capacity for low-carbon development in the rest of the world. As the living standards in China and other emerging economies or developing countries will rise dramatically, but these countries will not be willing or capable to chose anything else than the cheapest energy supply, buying down the cost of clean energy to normal market levels – the 'China price' – is the only way to prevent steep emission rises in these countries. Technological innovation in the US and elsewhere that lowers the price of low-carbon technologies can thus contribute to make clean energy options available and affordable for countries like China as well.

Friedman emphasizes that 'free-market capitalism' will have to do the job in the first place as '[t]he only thing as powerful as Mother Nature is Father Greed' (Friedman 2007). But governments could speed up the process and remove the uncertainties about the future price of carbon and the regulatory environment for new technologies through ambitious intervention in the form of efficiency standards, renewable quotas, cap-and-trade or a carbon tax.

Far more than a sober win-win story, Friedman describes the GND and building low-carbon economies as a panacea to the world's problems: '[P]residential candidates need to help Americans understand that green is about creating a new cornucopia of abundance for the next generation by inventing a whole new industry. It's [...] to preserve our American dream but also give us the technologies that billions of others need to realize their own dreams without destroying the planet' (ibid.).

GND policy proposals

It was the economic and financial crisis in 2007 that widened the public and policy debate on a Green New Deal and created both the space and the pressure for some form of political reaction. Most major economies economic reacted with some form of stimulus and reform programmes that usually included a green spending section. Building on this political momentum, various organisations took up the issue to make more substantial policy proposals for a Green New Deal.

It comes as no big surprise that the United Nations Environment Programme (UNEP) took a leading role among International Organisations in advocating a Green New Deal: the joint pursuit of economic prosperity and environmental integrity matches with UNEP's primary task, to promote sustainable development. Responding to the financial and economic crisis on the one hand and plans for national stimulus programmes on the other hand, UNEP called on G20 countries to dedicate a considerable share of their stimulus packages to green investment, and to invest at least 1% of their GDP for promoting green economic sectors (Sukhdev 2009).

In a more detailed policy proposal, UNEP suggests to combine short-term economic recovery with longer-term sustainable economic growth activities, 'an unprecedented global green deal of jobs, capital and technology flows to catalyze sustainable growth and avoid dangerous climate change' (UNEP 2009: 8). The long-term perspective requires more fundamental changes to domestic policies and the international governance architecture.

However, the GND debate remained superficial in most countries and focused on governmental stimulus packages to give a short-term push for economic recovery, rather than considering the long-term transformation of economies. Comparing national economic stimulus packages that governments put together as a reaction to the financial crisis, it was mainly the East Asian countries that allowed for a large share of green investment: while the Republic of Korea and China dedicated 79% and 34% of its total spending to green funds, this share was considerably lower in most other major economies (Sukhdev 2009).[83]

A prominent proposal for a more fundamental transformation process comes from the Green New Deal Group (GNDG), a loose coalition of British civil society and media professionals. The group was founded in 2007 to lobby for a government programme that addresses the 'triple crunch' of financial, climate and energy crisis: 'Drawing our inspiration from Franklin D. Roosevelt's courageous programme launched in the wake of the Great Crash of 1929, we believe that a positive course of action can pull the world back from economic and environmental meltdown' (Green New Deal Group 2008: 2).[84]

Contrary to other proposals, the group does not focus solely on a quick economic recovery that is compatible with environmental constraints; rather, it calls for a structural transformation of domestic and international economies and the financial system, not only to help foster investments in climate protection, but for general economic, social and ecological reasons. 'Finance will have to return to its role as servant, not master, of the global economy: to return to its given role of dealing prudently with people's savings and providing regular capital for productive and sustainable investment' (Green New Deal Group 2008: 23).

With respect to the climate crisis, restructuring economic and financial systems is of major importance to cut back the excessive influence of financial markets, and to enable national governments to implement effective economic policies and regulation. This includes the regulation of tax havens, reforms of taxations systems, the control of capital flows, global debt cancellation, and – last but not least – a paradigm shift away from credit-fuelled consumption towards a higher GDP share of public investment.

On the basis of this transformation process, the GNDG suggests a more immediate programme for economic recovery through investment in renewable energies and energy efficiency measures. Put another way, a fundamental reform of the way that financial markets work is seen as a precondition for any effective GND type of intervention.

The GND debate gained huge public attention and media coverage at the heights of the global economic and financial crisis. Given the massive concerns

---

[83] Australia 21%, France 18%, UK 17%, Germany 13%, US 12%, SA 11%, Mexico 10%, Canada 8%, Spain and Japan 6%.
[84] Members include the Economics Editor of the *Guardian* and former or current representatives of, *inter alia*, Greenpeace, Friends of the Earth, the British Green Party, and the New Economics Foundation.

for a global recession and its consequences, combined with the ecological modernisation dimension of these programmes, it comes as no surprise that proposals for a Green New Deal were generally well received in most countries. 'When paired with the inconvenient truths [...] of global warming, the idea of a Green New Deal is an easy next step for some constituencies to visualize necessary changes for the hard days ahead' (Luke 2009: 15).

UNEP further pushed the green growth agenda by introducing the notion of *green economy* as one of two central themes on the agenda for the United Nations Conference on Sustainable Development (UNCSD) 20th anniversary in Rio in mid-2012. The green economy concept further detaches the agenda from short-term economic recovery programmes, and turns to the benefits of greening economies and building green sectors and industries more in general. Initial discussions on the issue in (regional) preparatory meetings for Rio+20 support the perspective that building cleaner economies is widely seen as an opportunity for economic development and attracting higher levels of (foreign) investment, and a reform challenge for domestic institutions and frameworks in the first place. [85]

4.2.2 The low-carbon growth narrative

In a very similar way, low-carbon growth and development strategies focus on the domestic frameworks for economic transformation processes, which includes driving investment to the low-carbon activities of the future. Low-carbon growth strategies are a concern of public debates to a lesser degree than the GND issue, however, and widely remain within the expert circles of policy planning processes. Though addressing the same combination of economic and ecological concerns as the GND debate, these strategies take a long-term perspective on the transformation of economic organisation rather than focusing on immediate response measures, and they address the climate change challenge more explicitly as well.

---

[85] Though not all countries embraced the green economy concept in this way: in particular the Latin America regional group of countries rejected the approach altogether, while in the Asian and Arab preparatory meeting for Rio+20 countries tentatively welcomed it, and African countries broadly embraced the concept (see IISD coverage at http://www.iisd.ca/process/sustdevt.htm).

The common starting point for most low-carbon growth strategies is the increased attention to the magnitude of climate change and its consequences. During a plenary session at the 2010 World Future Energy Summit in Abu Dhabi, Nicholas Stern recounted that at the G8 meeting in Gleneagles in 2005, 'only two people were interested in climate change';[86] This had changed completely five years later when representatives from politics and business showed strong interest in the climate issue and, in particular, in low-carbon technologies. Nevertheless, Stern suggested, governments would address climate change much more decidedly if they would understand the full extent of the economic risks and opportunities related to it; an extent that he and his team underestimated as well when writing the Stern Review (Stern 2009a; Gordon 2011).

However, the magnitude of the challenge helped create an understanding that climate change cannot, like other environmental problems, be addressed through a few top-down regulations and adjustments, or an agreement on an international treaty mechanism; rather, it requires a fundamental shift in the way economies work and a profound socio-technical transformation (Newell and Paterson 2010).[87] 'Dealing effectively with climate change requires a dramatic restructuring of the global economy (because of the complete dependence of that economy on fossil fuel energy), making it incomparable with other environmental challenges. This claim is not, however, simply a normative exhortation; many political actors realise the necessity of this restructuring, and this realisation is increasingly widespread' (Paterson 2010: 363).

Pursuing low-carbon growth, or low-carbon development, is thus a way of aligning climate change concerns with economic objectives. But climate protection as low-carbon growth does not only require a new intensity of government intervention; it also entails a scalar shift. While the primary focus of climate change governance since the adoption of the Framework Convention in 1992 has been on the international negotiations and the success or failure of achieving a binding international agreement, low-carbon growth strategies turn the spotlight on domestic policies, in both developing and developed countries.

This can be interpreted in two ways: in a more optimistic reading, domestic climate regulation will complement international agreement and realise the commitments made in the negotiations. Many low-carbon growth strategies explicitly start from actual or intended domestic emission reduction targets and consider ways to implement these on the ground (see below).

---

[86] Own documentation.
[87] Interview Rupert Edwards, Climate Change Capital.

The more pessimistic story is that a bottom-up approach to climate change will increasingly replace the – thus far – resultless quest for an ambitious global climate deal, reducing the international dimension of climate politics to the coordination of national efforts and exchange on best practices.[88]

A related argument is that the magnitude of the climate change challenge and the need for a fundamental transformation of economies means that the issue has become too big for the institutions of global climate governance, or vice versa, that the UNFCCC mandate is too limited to organise a global economic transformation. Climate change as an economic challenge would have to be addressed within the international governance arenas that are concerned with the organisation of economic processes more immediately.[89]

A bottom-up approach that focuses on domestic transformation processes is thus often described as a more realistic approach to climate change governance. In that sense, the objective is 'to move the debate beyond [...] the likelihood of bargaining trade-offs in and detailed design of an international emission reductions agreement' (Policy Network 2009: 11), and to focus on stimulating domestic action instead. Low-carbon growth strategies, in that sense, can be understood as developing the domestic climate change institutions and policies that are still lacking (Giddens 2009).

Low-carbon strategies seek to fill the gap of domestic climate regulation by formulating policies and programmes for long-term emission reductions that are consistent with economic growth and competitiveness priorities. A primary focus in these strategies is usually on building cleaner energy systems, in both developed and developing countries: 'The only chance of slowing the buildup of CO2 concentrations soon enough to avoid catastrophic climate change that could take centuries to reverse is to transform the energy economies of industrial and developing countries almost simultaneously' (Flavin 2009: 5).

The strategic side: highlighting opportunities

The low-carbon growth narrative is sustained and strengthened by a sense of the economic opportunities that arise from restructuring economies, in the form of immediate gains through enhanced resource efficiency or from higher competitiveness in a climate constrained future. In its general form, the idea has been part of environmental discourses for at least two decades, and it is at the heart of ecological modernisation strategies that emphasize the economic benefits of resource efficient economies.

---

[88] Interview Tom Heller, Climate Policy Initiative.
[89] Interview Kirsty Hamilton, Chatham House.

The low-carbon development or growth narrative itself has various starting points. The McKinsey cost curve and similar studies played an important role in highlighting the opportunities to expand economic product while reducing climate risk. In that sense, climate protection as low-carbon development is about realising productivity gains in those areas where energy savings through low-emission production or consumption result in financial savings as well. A great number of studies seek to prove that this is often the case for enhancing energy efficiency and other forms of resource efficiency, while renewable energy investments would take more time to write off.

The importance of making these opportunities visible and almost palpable is reflected in the way that cost curves have spread as a crucial tool in climate policy-making. Project Catalyst, the think tank that commissioned and sponsored the original McKinsey analysis, advises many governments on cost curve issues and low-carbon development plans.[90] The cost curve approach is also fundamental for other organisations like the World Bank in supporting governments with the low-carbon transformation (Johnson, Alatorre et al. 2010).

In many cases, however, low-carbon gains still exist as an idea only. In that sense, focusing on the opportunities and benefits of emission reductions is a strategic approach to make climate protection more feasible, as it 'became apparent that politically, as long as climate risk management was understood as a constraint on growth, a trade off with growth, that invariably it was going to lose in that trade off'.[91] Framed as the opportunity to develop new technologies and economic activities instead of a burdensome task, climate protection is more likely to gain support. Moving to a low-carbon future, in that sense, 'requires a strong political narrative of hope and opportunity' (Policy Network 2009: 13).

Low-carbon growth, then, is a positive story that allows describing climate protection as a benefit to all. The frequent efforts to communicate these opportunities show, however, that the economic opportunities of climate protection are still to emerge in many cases and are not always easy to realise. As former US President Bill Clinton announced in the run up to the 2009 climate negotiations: 'I just hope that the people in Copenhagen won't lose sight of the fact that there are economic opportunities out there. [...] This is being sold as a dose of castor oil you have to swallow and it's just not true.'[92]

At the climate talks in Cancún one year later, German environment minister Norbert Röttgen encouraged delegates to invest in climate protection, arguing that these investments pay off in the long run: 'In business, politics and society,'

---

[90] Interview Ramzi Elias, Project Catalyst.
[91] Interview Tom Heller, Climate Policy Initiative.
[92] http://www.reuters.com/article/2009/12/08/us-climate-copenhagen-billclinton-idUSTRE5B64MQ2 0091208.

he suggested, 'we no longer see climate change as a threat, but an opportunity and a challenge'.[93]

For those who push climate protection as a positive story this way it was disappointing to see that the climate negotiations in Copenhagen 2009 'were still about burden sharing',[94] that is, dominated by the idea that reducing emissions entails costs rather then opportunities.

It is these competing perspectives on the climate issue that Nick Robins from the UK based bank HSBC found himself confronted with when he joined the climate talks in Copenhagen with a group of investors. Eager to sign low-carbon deals, they ended up facing pessimist government talks rather than the positive vibe of an investment fare: 'I got the impression that there were lots of sort of very confused business people and investors, wandering around, saying: look, we are here for this opportunity, where is this? They were expecting a sort of party, a low-carbon party, where in the end, people weren't really talking about this'.[95]

4.2.3   Low-carbon growth and development: country perspectives

Low-carbon growth or development strategies often build on this sense of opportunity as an important argument to engage stakeholders for the low-carbon transformation. The following looks at exemplary country cases to describe how their low-carbon strategies evolved, and whether and how they address the need to govern investment.

It is not possible, however, to draw a complete picture of approaches to low-carbon growth, due to the evolving nature and diversity of these strategies in different world regions, not least regarding the degree to which governments formulate respective strategies. Describing examples of low-carbon strategies for developed countries (UK), emerging economies (Korea, China), and developing countries (the creation of NAMAs as a way to support low-carbon development) is therefore not meant to compare different approaches, but rather to add different pieces to the evolving low-carbon picture, and to ask for the common ground on which low-carbon strategies stand.

---

[93] http://www.dw-world.de/dw/article/0,,6311489,00.html.
[94] Interview Ramzi Elias, Project Catalyst.
[95] Interview Nick Robins, HSBC.

Developed countries: the United Kingdom and the EU

Though the UK has not been a front runner in achieving emission reduction always, it is an exemplary case in the low-carbon debate as it was among the first industrial countries to explicitly address the transformation to a low-carbon economy. The British government made an early intervention with its 2003 Energy White Paper, *Our energy future – creating a low-carbon economy*. Climate change concerns are supplemented here with a sense for the economic opportunities of fast mover advantages in developing low-carbon technology. As then Prime Minister Tony Blair put it: 'As we move to a new, low carbon economy, there are major opportunities for our businesses to become world leaders in the technologies we will need for the future' (UK Department of Trade and Industry 2003: 3).

The low-carbon growth issue gained increasing importance in the UK a few years later and was taken up by several government bodies. The general objective is to embark on an economic growth trajectory that is consistent with cutting emissions by at least 80% below 1990 levels by 2050. Building on UK and EU emissions targets and a bottom-up sector by sector analysis of feasible emissions reductions and likely costs, the Climate Change Committee (a scientific advisory body established by the Climate Change Act) recommends a 42% emissions reduction in case a global climate deal is sealed, or a 34% reduction without such a deal, highlighting 'that reductions of that size are possible without sacrificing the benefits of economic growth and rising prosperity' (UK Committee on Climate Change 2008: 1).[96] However, costs will only be that low 'with appropriate policies and given early action to put the UK on an appropriate path' (2).

This early action is meant to secure that the majority of emissions cuts happens domestically: emissions trading is meant to play an important role in reducing the total cost of emissions cuts and in providing financial flows to developing countries, but 'it is essential for rich developed countries to achieve significant domestic reductions to drive the development of required low-carbon technologies' (9). The Committee emphasizes that building a low-carbon economy is less a technical challenge and rather an issue of political will and 'strong leadership from government' (1).

Though emphasizing the importance of early action and an active government role, the report is rather indefinite on what this action should look like. In a contribution for a book published by the Labour-related think tank Policy Network, Peter Mandelson, then UK Secretary of State for Business, Enterprise and

---

[96] The Committee estimates that the costs of these reductions will be between 1-2% of GDP in 2050, and thus at 'the same order of magnitude as those provided by the Stern Review and other global and UK studies' (2).

Regulatory Reform, makes a number of more concrete proposals (Mandelson 2009). Describing the challenge as 'a problem for industrial policy in the broadest sense' (91) he suggests that the focus must be on the immediate economic benefits of low-carbon development, and thus on the areas where companies are able to capture first mover benefits. The government should send 'clear, consistent and unambiguous signals' (90) for investment decisions through a coherent and stable regulatory framework, including targets for emissions levels and renewable energy production.

Additionally, he suggests the development of infrastructure for large-scale renewable and local energy production, an increase of government support for R&D in clean technologies, and a secure and steady flow of venture capital to low-carbon firms, in particular to small companies and projects. Though ambitious emissions reductions should drive the demand for green technologies, it might be necessary to enhance demand through targeted loans and subsidies for implementing energy efficiency in both the public and private sectors. Altogether, this 'activist industry policy' (95) should give confidence to business and investors, supplement the market through a price on carbon and infrastructure development and support British companies to become fit for international competition.

In 2009, the Departments for Business, Innovation and Skills, and Energy and Climate Change jointly published *The UK Low Carbon Industrial Strategy* that has close ties to the publication of the Stern Review.[97] The respective Secretaries of State Peter Mandelson and Ed Miliband emphasize that tackling climate change 'can create a better kind of society and a stronger, more sustainable economy' (UK Government 2009: 4); they highlight economic opportunities in technology development and manufacturing, and in additional sectors such as green consultancy or low-carbon venture capital services.

The report emphasizes that 'the combination of both the massive dynamism of the private sector and a strategic role for government' (5) is required to obtain the maximum benefits from the transition to a low-carbon economy: governments must take a long-term and stable strategic approach and a 'pragmatic' (6) approach to markets, and ensure that British companies are ready for the competition and demand created by global climate change policies. Among the concrete measures for a low-carbon transformation that strengthens domestic markets is

---

[97] The Commission on Environmental Markets and Economic Performance (CEMEP) that was established to consider the significance of the Stern Review results for the British economy in 2007 published a report with recommendations for the government and businesses on how to stimulate investment in environmental markets (BERR, Defra et al. 2007). The government response to this report (UK Government 2008) is the basis for the Low Carbon Industrial Strategy.

early stage innovation support, an Innovation Investment Fund, and interest free loans for investment in energy and resource efficiency.

A report by the *Green Investment Bank Commission*, an advisory council with members from the finance industry[98] convened by the Chancellor of the Exchequer, specifies the challenge of *Unlocking investment to deliver Britain's low carbon future* (GIB 2010). It focuses on the market failures and barriers that prevent greater investment to low-carbon technologies, including: capacity limits of investment markets, as low-carbon investments are often perceived as riskier and thus are 'unlikely under current market conditions to attract the capital needed within the time desired' (5); regulatory risks, as the return on low-carbon investments strongly depends on government incentives and regulation; and the need for large numbers of small investments that challenge the current organisation of investment markets.

The report argues that government intervention is required to overcome these barriers and proposes as one crucial instrument a state owned Green Investment Bank (GIB), that aims at stimulating investment flows through risk mitigation 'rather than simply increasing rewards to investors' (xiii). The bank should deploy public money within public-private investments that address specific market failures and achieve lowest-cost emissions reductions. The GIB is meant to be commercially independent and not accountable to the government or Parliament for individual investment and lending decisions, as a necessary prerequisite for building credibility with markets. The GIB governance structure, then, 'should clearly manage the tension between investing in the public interest and the need to be commercial' (xv).

The UK low-carbon growth strategy shows that various ministries are involved in planning this transformation, with a leading role of the economic and financial bureaucracy. It also highlights the transformational dimension of building low-carbon economies, and the active role of governments in bringing about this transformation – a role that is increasingly accepted and embraced within business and the finance industry as well (see section 4.3).

The UK perspective is supported by the European Commission that more recently published a *Roadmap for moving to a competitive low carbon economy in 2050*, asking for cost-effective ways to achieve the required GHG reductions (European Commission 2011). The roadmap draws strongly on long-term modelling to formulate equally long lasting policy proposals. The fundamental challenge for building the competitive economies of the future, the roadmap suggests, is to drive investments to future technologies at an early stage. Andris

---

[98] Yell Group, CC Capital Generation Investment Management, Alliance Trust, Logica plc, E3G, Citibank, Bank of America Merrill Lynch, Foresee Ltd, Carlyle Group, Hg Capital, Oliver Wyman, Goldman Sachs.

Piebalgs, European Commissioner for Energy, describes the quest for Europe's Energy Future as a New Industrial Revolution: 'Europe has the chance to establish world leadership in clean, efficient and low-emission energy technologies. These will become an engine for growth and job creation whilst sustaining a high quality of life' (Piebalgs 2008).

Emerging economies

However, the push for building low-carbon economies does not exclusively – or not even in the first place – come from those western industrial countries that led much of the economic and technological development of the 20th century. It is the energy policies and markets of Asian countries, in particular China and India, that have 'begun to change rapidly – more rapidly than those in many industrial countries' (Flavin 2009: 5).

As we have seen above, it was China and Korea who had the largest green economic stimulus programmes, either in absolute or in relative terms. It is these Asian countries as well, according to analysts like Tom Heller from the Climate Policy Institute, that strongly contributed to developing the low-carbon growth perspective, visible in particular in the Korean green growth strategy.[99]

The pioneer role of Asian countries is explained by some through the fact that building low-carbon economies is essentially an industrial policy approach to climate change, an approach for which western industrial countries no longer serve as a model, and which fits much better with economic policy in many Asian countries: 'We see some of the most interesting activities in some of these countries that still have relatively strong institutional residues of industrial policy'.[100] Put another way, these countries have 'much more of a planned mentality to economic growth', which includes the belief that forcing money into low-carbon technologies will lead to economic growth – a belief that is less developed in European countries, as Nick Robins observed.[101]

The two paradigmatic cases for low-carbon growth strategies in Asia are Korea and China: while Korea has announced ambitious plans for green growth and is praised by many, including UNEP, for its efforts in reducing GHGs, China seems to be less explicit about its low-carbon and climate change strategy, but nevertheless addresses the same issues through its energy legislation.

Korea's *National Strategy and Five-Year Plan for Green Growth* aims at transforming the country's growth paradigm from *quantitative* to *low-carbon or*

---

[99] Interview Tom Heller, Climate Policy Initiative.
[100] ibid.
[101] Interview Nick Robins, HSBC.

qualitative growth (Presidential Committee on Green Growth 2009). The green growth plan announced in 2009 revives the tradition of regular five-year economic development plans that Korea has implemented since the 1960s and that 'had been very effective during the early development era of the Korean economy' (11). This practice was abandoned during the 1990s as the effectiveness of planning was seen to decrease and the Korean economy embraced market economy principles. Returning to five-year planning does not mean, as the government emphasizes, that Korea is returning to the development era; rather, these plans are a 'comprehensive long-term master plan' (8) to build consensus on the required changes and give a clear signal for private sector investment and activities.

The green growth strategy follows the three main objectives to mitigate climate change, to create new engines for economic growth, and to enhance the quality of life. Beyond green industrial policy, it includes programmes for green cities and educational programmes on green growth. The centrepiece of the strategy, however, is the objective to spend 2% of GDP on green growth, increasing both public and private investment to green sectors, such as clean technologies and renewable energy, resource and material efficiency, sustainable agriculture, green construction, sustainable cities and transport, and ecosystem restoration.

The first component is a green stimulus package that is part of Korea's Green New Deal programme, in the form of financial, fiscal and taxation policies that account for 4% of Korea's GDP in the period from 2009-2012; a large share of these initial investments goes to infrastructure development, but expenses for research and development are expected to rise quickly. The 2009-2013 five-year plan, as the second component of the green growth strategy, provides incentives for private (green) investment, including tax benefits, long-term and low-interest green bonds, easy credit access, credit guarantees, and a green private equity fund.

The Korean Green Growth strategy is of particular interest as it is the first and most far-reaching attempt in this direction, and consequently attracts the interest of many stakeholders. The UN Environmental Programme (UNEP) praises the efforts made by the Korean government thus far, but cautions that targeted investment will not suffice to enable the emergence of a green economy, and urges for concomitant policy reforms on both the domestic and international level, in particular the reform of subsidies, pricing systems and taxation that encourage harmful resource extraction and pollution in the energy sector and beyond (UNEP 2010). Supporting these findings, an OECD study additionally calls for a market approach to reducing carbon emissions through an emissions trading system supplemented by a carbon tax in the areas not covered by the trading system, and for spending public expenditures in a way that does not 'pick winners' (Jones and Yoo 2010: 23).

The NGO network *Green Korea United*, to the contrary, rejects the Korean Green Growth strategy as 'no more than an economic development vision' driven by large-scale civil engineering projects and constructions, and criticises in particular the 'shamefully low GHG reduction target' that would not meet the country's historical responsibility for climate change and its capacity for emissions reductions, and the projected increases in energy use (Green Korea United 2009).

China is a particular case with respect to orchestrating the low-carbon transformation. Opinions vary widely on the achievements and appropriateness of the country's efforts in reducing or stabilising emissions, and this variation is attributable to political interests in many cases (The Brookings Institution 2011). What becomes obvious, however, is that the Chinese government is less explicit in addressing low-carbon growth as a strategy to reduce emissions and combat climate change, but rather aims at energy and resource savings through its legislation and recent Five-Year Plans.

In 2005, China introduced a Renewable Energy Law and set Renewable Energy Targets on a par with those of the 27 EU member states, aiming for 15% of total primary energy consumption from renewable energies by 2020 (Urban and Yu 2009). In its 11th Five-Year Plan, China formulated the goal to reduce the energy intensity of its economy by 20%. A study by the Climate Policy Initiative concludes that this target has been widely met; the authors identify a significant increase in regulatory and policy activities for lowering emissions, and ascribe at least parts of the lower energy intensity to these efforts (Climate Policy Initiative 2011).

Likewise, the Chinese government gives increasing emphasis to creating a low-carbon and green economy, and is considering policy instruments other than top-down regulation, including emissions trading and carbon taxes: the most effective policies for China were top-down administrative measures, according to Qi Ye, Director of the Climate Policy Initiative, but often these measures were very expensive as well: 'the policies in the 11th Five-Year Plan were effective but not efficient'.[102]

It is often suggested that these initiatives are driven by domestic concerns rather than global climate change objectives, including energy security, local environmental pollution and supply diversification. It is in particular the reversal of the long-term decline in energy intensity in the years 2002-2005 that is said to concern Chinese leaders, but also the acceleration of GDP growth and the growth

---

[102] Qi Ye, Director, The Climate Policy Initiative/Tsinghua University (cf. The Brookings Institution 2011).

of energy intensive industries, and sharply increasing energy prices (Climate Policy Initiative 2011).[103] Rather than climate change concerns, the reason for enhancing resource efficiency seems to be concern for resource constraints to China's growth path, though it is sometimes claimed as well that the Chinese government is increasingly worried about being blamed as the world's largest greenhouse gas emitter (ibid.).

Rather than resource conservation at the expense of economic growth, the Chinese strategy is to massively scale-up its capacity in renewable energies, with effects that go beyond the Chinese economy: 'It's just a dramatic expansion of capacity in a very short period of time, but even if they get halfway there, this will transform, fundamentally, the global market for clean energy technology, to be sure. It will change its price points. It will change the relative economics of low-carbon technology versus high carbon technology'.[104]

Low-carbon development – NAMAs

Finally, the low-carbon discourse has a strong development trajectory as well: given the growing consensus that developing countries will increasingly have to contribute to global emissions reductions (or at least limit the growth of emissions) but are anxious to curb economic growth, low-carbon development is seen as a way of aligning domestic economic interests with global needs. It is closely related to the climate finance debate in this respect, as most developing countries will be neither willing nor capable of financing the more expensive clean development trajectories, and thus require external financial support. Supporting low-carbon growth this way is seen as a promising way of using climate finance funds to leverage much larger private investments (WEF 2011).

The National Appropriate Mitigation Actions (NAMAs) concept, introduced by the Bali Action Plan and specified by the Copenhagen Accord, allow developing countries to realise mitigation actions in the framework and as part of their wider development strategies. NAMAs are entitled to financial and technical support and subject to international monitoring and verification. Whereas the Bali Action Plan refers to both developing and developed countries,

---

[103] Opinions vary on how much of the decreases in energy intensity over several decades are due to regulation and policy incentives, or to the contrary, 'primarily the result of Beijing letting go of the economy, getting out of the way a little bit. [...] What we saw from 1978 to 2002 was a dramatic decline in energy intensity of the Chinese economy because people went from making steel with no concern about energy costs, because prices weren't liberalized and they didn't have a profit incentive, to making Barbie dolls and DVD players' (Trevor Houser, The Rhodium Group, cf. The Brookings Institution 2011: 17).
[104] Trevor Houser, The Rhodium Group cf. The Brookings Institution 2011.

suggesting *nationally appropriate mitigation commitments* by all developed countries and *nationally appropriate mitigation actions* by developing country Parties, the Copenhagen Accord uses the NAMA concept for Non-Annex 1 countries exclusively.[105]

But while the Accord specifies that NAMAs will be recorded in a registry together with relevant technology, finance and capacity building support and subject to international measurement, reporting and verification, it is less specific on the form of NAMAs; it explains that Non-Annex I Parties 'will implement mitigation actions [...] in the context of sustainable development' (UNFCCC 2009: 2). At COP 16 in Cancún in 2010, developing countries once more supported to undertake NAMAs as a way of reducing expected business as usual emissions by 2020 (Climate Focus 2011).

Financing mitigation in developing countries as part of low-carbon development strategies through NAMAs changes the governance of (climate) finance flows in several ways. First, funding for emissions reductions in developing countries is no longer based, as was in the case of the Clean Development Mechanism, on a global carbon price and the objective of global cost-effectiveness achieved by a global carbon market; rather, it can support various forms of mitigation in developing countries that do not compete with each other through the market coordination. NAMAs have therefore been criticised as 'enormous and undisciplined subsidy programs' that lead away from carbon pricing (Stoft 2009: Preface).

Second, supporting NAMAs in the context of low-carbon development at the same time replaces the idea of publicly funded mitigation and adaptation measures in developing countries, towards a broader notion of financing abatement: 'It is argued that it would be better to move away from thinking of solutions in terms of "least-cost mitigation + adaptation", to thinking of it in terms of "low-carbon economy + development". Such a move permits the involvement of different political and economic actors (in particular, the private sector) and relevant institutions, and allows for greater cooperation, decentralization and competition' (Zadek 2011).

What follows from framing support for mitigation activities in developing countries in this way is that neither the carbon market nor public funding alone are expected to deliver the necessary finance; rather, both are part of, and instruments for, encouraging investment for low-carbon development.

---

[105] Different understandings prevailed, however, as India has argued that NAMAs refer to *voluntary reductions* by developing countries only, and which are required to be supported and enabled by technology transfer from developed countries: http://unfccc.int/files/kyoto_protocol/application/pdf/india100209b.pdf.

The primary focus is then on the domestic reforms and frameworks as a crucial precondition for enabling low-carbon investment.[106] NAMAs are crucial, in that sense, in helping to finance these frameworks and support capacity and institutions building. The public money channelled through NAMAs would cover the incremental costs for investors relative to alternative, high carbon technologies, and is expected to play an essential role in mobilising private investment this way (Project Catalyst 2009b). NAMAs are an important tool within low-carbon development plans as they help to distinguish between domestically and internationally financed actions, and thus implement the idea that international support should only cover the costs additional to BAU development.[107]

Most low-carbon development plans were still in the making when this book was written. Guyana was among the first countries to formulate a strategy to align national development needs with a contribution to climate protection: 'As a country [...] we wanted to break the false debate which suggests that a nation must choose between national development and combating climate change' (Republic of Guyana - Office of the President 2010: 5). The objective of the strategy that is praised by the AGF as an 'Example of spending wisely' (AGF 2010: 53) is to create a low-deforestation, low-carbon, climate resilient economy, taking Guyana's National Development Strategy (NDS) and the National Competitiveness Strategy (NCS) as starting points.

Guyana's main contribution to reducing global emissions could be in lowering its deforestation rate. The core challenge, the development strategy explains, is in reconciling the 'tension between protecting rainforests and pursuing economically rational development' (7). A forest finance mechanism such as REDD(+), in that sense, 'would make forest protection an economically rational choice' (ibid.).[108] Guyana supports a phased approach to REDD+ that starts with a fund-based mechanism for REDD+ from 2010, to gradually merge into the carbon market.

The REDD+ money would not only finance forestry policies and projects, however, but be used for investment in Guyana's overall low-carbon development, and to ultimately increase private investment in other (low-carbon) economic sectors and infrastructure as well. Attracting large-scale catalytic investors to Guyana will, according to the government, 'require incentives to finance in-

---

[106] Interview Ramzi Elias, Project Catalyst.
[107] ibid.
[108] Based on an assessment by McKinsey, the Office of the President estimates the value of Guyana's rainforest, if harvested and the land put to the highest value subsequent use, to be between US$4.3 billion and $23.4 billion.

dustry-specific infrastructure and overcome perceived country investment risk'
(12).

4.2.4 Going low-carbon: towards economic rationality

The Green New Deal and low-carbon discourses raise the awareness of the need
to decarbonise economies, with respect to possible constraints through climate
change legislation, but increasingly also as a factor of economic competitiveness.
The economic stimulus programmes after the financial crisis pushed this understanding. Ban Ki-moon, UN Secretary General, urged that these programmes
'must help catapult the world economy into the 21st century, not perpetuate the
dying industries and bad habits of yesteryear'.[109]

But greening the economy has become an imperative beyond these recovery
programmes as well. A report from Deutsche Bank Group's Climate Change
Advisors suggests that first movers to a low-carbon economy will increase their
growth rates, and sees countries like Germany and China well positioned due to
their emission reduction and climate change policies.[110] To quote Ban Ki-moon
once more: 'Leaders everywhere [...] are realising that green is not an option but
a necessity for recharging their economies and creating jobs'.[111]

Debates on the low-carbon transformation within the US in particular highlight the complex relationship between climate protection objectives and economic imperatives in going green – and how this is related to international competitiveness: in the US, Trevor Houser suggests, 'conversations about China's
energy policy or action or inaction on climate change generally happens in fairly
cartoonish ways driven by people's domestic political priorities, so if you are
interested in having the U.S. pass climate change legislation or deploy clean
energy, then China is winning the race in clean energy and is about to wipe the
U.S. off the map and we're seeing our economic future vanish in front of us. And
if you are not interested in seeing the U.S. pass climate change legislation, then
it's all smoke and mirrors in China' (cf. The Brookings Institution 2011: 15).

Even if China would pursue higher energy intensity and the extension of renewable energy supply for purely domestic and economic reasons, this would
have an effect on the situation of other countries, as China is increasingly seen as

---

[109] Ban Ki-moon: Green growth is essential to any stimulus. *Financial Times* (London), 17 February
2009.
[110] Reuters 2010: China, Germany lead the race toward a low-carbon economy. Mar 25, 2010,
http://www.reuters.com/article/2010/03/25/us-solveclimate-china-idUSTRE62O3PL20100325
[111] Ban Ki-moon: Green growth is essential to any stimulus. *Financial Times* (London), 17 February
2009.

a rival for higher shares in the emerging low-carbon technology and renewable energy markets. Going low-carbon is thus economically rational from a global point of view, as the Stern Review had argued, but also from a domestic perspective, an argument that is increasingly voiced in the US as well: 'And that's where the challenge is: if we don't want to be left behind, we will have to do something similar.'[112]

The Green New Deal and low-carbon growth discourses thus open a space for new government practices, and for reconsidering the role and responsibility of governments in the organisation and strategic development of economies. Clearly, approaches to realising the low-carbon future still vary widely, between targeted sectoral policies to stimulate investment and more radical transformation programmes.

These differences reflect, to a certain degree, the openness of this process as well: while the low-carbon growth discourse is sustained by a strong sense of opportunity, the exact way for realising these opportunities is subject to a search process on the one hand, and struggles between different interest groups on the other hand: while some sectors and thus groups within society can expect long-term benefits from the low-carbon transformation, the short-term disruption of other sectors will also create losers from the transformation process.[113]

This once more highlights that orchestrating the low-carbon transformation is not about building a green parallel universe within the economy, but about making low-carbon the very heart of economic development and future prosperity. There is a great consensus in the low-carbon growth discourse that incentives and regulation have to play an increasingly important role in stimulating investment to the sectors, industries and technologies that will form the green economies of the future.

Organised business and finance groups also broadly support a more active approach to enabling and stimulating investment in low-carbon sectors and industries. The World Business Council on Sustainable Development (WBCSD) calls for policy instruments that reduce investment risks, price carbon and increase the return on low-carbon investment (WBCSD 2011). A future climate policy framework should strive to unleash large-scale investment by addressing investment barriers and using public funding to leverage private finance (WBCSD 2009). The WBCSD also supports NAMAs in developing countries in the form of combined local, national and sector programmes that would be most effective in aligning economic growth and emissions reductions (WBCSD 2011).

---

[112] Joe Klein: What a New Energy Economy Might Look Like. *Time*, 13.11.2008: http://www.time.com/time/politics/article/0,8599,1858684,00.html.
[113] Interview Tom Heller, Climate Policy Initiative.

The World Economic Forum's Task Force on Low-Carbon Prosperity likewise supports government intervention this way, but cautions that achieving a low-carbon transformation is not a task for governments alone. 'As the key delivery agent of low-carbon investment, innovation, products and services, business needs to have a voice at the table'.[114] The next section will take a closer look at what the business and the finance industry has to say about the appropriate policy frameworks for building low-carbon economies.

4.3 The finance sector in climate change governance

Visitors to the 2010 World Future Energy Summit in Abu Dhabi could have two very different impressions of the current state of the low-carbon transformation. In the spacious pavilions of the super-modern Abu Dhabi National Exhibition Centre, this transformation seemed to be in full swing, as investors strolled along the booths to learn about clean energy start-ups and the low-carbon innovations of big transnational companies, looking for investment opportunities in clean technology and renewable energy projects. The low-carbon transformation, it seems, is already driven by economic interests and opportunities.

At the policy part of the Summit, to the contrary, it appeared that the transformation to a low-carbon future is under heavy construction, and that the panels and workshops of the Summit are among the places to create the most effective tools to make this transformation happen: government officials from many countries and members of the different shades of the business and investment communities discussed the policies and measures that can help to kick-start the next industrial revolution.

These debates show that investors have a very specific perspective on the low-carbon transformation, and on the risks and opportunities of financing clean technologies and renewable energies; this section takes a closer look at this investor perspective. It starts with the early attempts of a few (international) organisations to increase the interest of the finance sector in financing renewable energy development. This work receives more attention within climate change governance more recently, due to the interest in low-carbon strategies and in the deployment of renewable energy and clean technologies.

Increasingly, investor groups participate in climate change debates as well: It is important to see that this is more than the lobby effort of a special interest group and rather sustained by the interest of government agencies, IGOs and consultancies in the perspective of financiers. The investor perspective on cli-

---

[114] World Economic Forum Task Force on Low-Carbon Prosperity, CEO Climate Policy Recommendations: http://www.weforum.org/issues/climate-change.

mate change governance is thus crucial for how investment for climate protection is made an object of government.

4.3.1 Renewable energy governance: Bringing in the finance sector

The first attempts to engage the investment and finance community as a crucial actor for climate protection started in the early 1990s, with initiatives by the International Energy Agency, Greenpeace, and the UNEP Finance initiative. At the beginning of the 2000s, global renewable energy governance was established through events like the 2002 World Summit on Sustainable Development in Johannesburg and the Renewable conference in 2004 in Bonn; since 2009, the International Renewable Energy Agency (IRENA) with almost 150 member states (November 2011) promotes the global uptake of renewable energies.

While activities to support investments in renewable energies steadily increased since the mid 1990s, international climate policy considered these sectoral measures to a much lesser degree: the UNFCCC frames climate protection as emissions reductions, and the market and flexibility approach of the Kyoto Protocol affirms that it matters less in what sector or world region emissions reductions take place, and rather that these reductions occurs in a cost-effective way. Renewable energies are only one among several options for emissions reductions, and less likely to be chosen due to this least cost approach to climate protection.

Additionally, the private sector opinion in climate politics was largely voiced by incumbent industries for long, often related to the fossil fuel sector, and thus disapproved any form of tighter climate regulation. It was not least the lobbying power of what was often perceived to be *the* private sector that prevented ambitious climate protection measures, on both national and international levels. The fossil fuel industry is seen as crucial for the non-ratification of the Kyoto Protocol through the US; and in the European Union, industry organisations achieved a generous assignation of allowances in the EU Emission Trading Schemes (ETS) that watered down the system's effectiveness (Brunnengräber 2007).

More recently, however, awareness is rising that things look very different from a finance sector perspective. Christiana Figueres, who was appointed UNFCCC Secretary General in 2010, is among those who emphasize that financiers would be happy to shift their investment portfolios into renewable energies and clean technologies – if stable and reliable investment frameworks would be in place to support this shift.[115]

As this argument is not new but has been repeatedly stressed for almost two decades, the increasing attention for an investor perspective must be explained within a greater shift in the approach to climate change: while the understanding of the economic risks and opportunities related to climate change has generally initiated a search for new strategies to enhance climate protection while achieving economic objectives, it is in particular the climate finance and low-carbon growth discourses that highlight the need for stimulating and regulating investment for climate protection in both developed and developing countries.

Greenpeace and the insurance industry

One of the first instances of a non-profit organisation engaging the financial sector in climate change governance was the cooperation that Greenpeace sought with insurance companies in the early 1990s. Sustained by the effort and enthusiasm of Jeremy Leggett, then the Scientific Director of Greenpeace's Climate Initiative, the campaign aimed at voicing the perspective of those businesses that have an interest in tighter climate change regulation.

While this would have applied to various sectors (Greenpeace considered the tourism, agriculture and water industries as potential allies which are all vulnerable to climate change), the organisation decided that the insurance and banking community was most likely to be a partner for supporting ambitious climate protection (Hohnen 1999).

The insurance sector, and in particular reinsurance companies, expected higher risks from climate change than other businesses, as it would have to pay for the losses from – increasingly climate driven – extreme weather events and other natural disasters.

Apparently, insurance businesses were aware of climate risks very early on: in the late 1980s, large reinsurance companies created their own climate research programmes and cautioned against climate risks, alarmed by a series of extreme weather events throughout the 1980s; as early as 1990, the General Manager of Swiss Re suggested that the increasing number of natural catastrophes that caused unprecedented losses for insurance companies are probably related to the effects of climate change;[116] and after the Earth Conference in Rio in 1992, Gerhard Berz from Munich Re wrote: 'The insurance industry can only hope and wish, in its own interest, that the convention to protect the world's climate [...]

---

[115] Personal communication: UNFCCC Climate Talks, Bonn, June 2006.
[116] http://archive.greenpeace.org/climate/database/records/zgpz0678.html.

will help achieve a political breakthrough for the many good plans that have been proposed and succeeds in healing the earth from the "fever" which is beginning to afflict it' (Legget 1993: 46).

Jeremy Leggett and his colleagues at Greenpeace hoped that the growing awareness of the vulnerability to climate change would cause trouble for insurance companies as they often invested large sums in oil and coal companies and thus financed economic activities which, in the medium and long run, could lower the profitability of their core business. Shifting these investments to renewable energies, Leggett suggested, could make a big contribution to the development of the renewable energy industry at the expense of the incumbent fossil fuel industry: 'The insurance industry collects some $1.4 trillion in premiums every year. [...] A lot of it goes to fossil fuels, which only makes things worse, and almost none to solar and other renewables. We'd like to reverse that' (cf.: Hertsgaard 1998: 268).

When Greenpeace started approaching the sector in the early 1990s, it found that insurance companies were aware of the risks they faced and happy to exchange insights on these issues, but rather reluctant to take a public profile on the climate issue (Hohnen 1999). The NGO therefore tried to win the insurance industry as a strategic partner that, beyond studying and calculating climate risks for their investment portfolios, takes an active and visible role in the climate negotiations. In the person of Jeremy Leggett, Greenpeace started to network and exchange opinions with some of the world's leading reinsurance companies, including Munich Re, Swiss Re and Lloyds of London.

In the coming years, some of these companies began raising their voice through media releases or expert publications. Jeremy Leggett made various efforts to support these statements, including a workshop with members from the insurance, reinsurance and banking industry during COP 1 in Berlin in 1995 (Legget 2001). Again later, Greenpeace extended its networking to the solar energy industry, the sector that was meant to receive much of the investment flows diverted away from the polluting industries.

However, opinions vary whether the Greenpeace initiative was a success, and ultimately had an impact on the behaviour of (at least parts of) the financial industry. Pearce suggests that Greenpeace had 'shown the way' by engaging the insurance companies in the international climate change negotiations, and had achieved a 'coup' by persuading them to speak out (Pearce 1997). Hohnen, to the contrary, 'lays no claim that the work described had any more than a catalytic effect, and was no more than one of many factors which eventually encouraged the insurance and banking sector to make public statements supporting government action to protect the planet from human-induced climate change' (Hohnen 1999: 3).

Jeremy Leggett himself is rather sceptical regarding the achievements of his campaign, in particular with respect to the readiness to change behaviour within the insurance industry. He notes that a significantly larger number of insurance companies had come to the Kyoto COP in 1997 than to previous climate meetings, but laments that 'it became clear that people within the threatened industry itself still didn't seem to grasp the reality. [...] The sad truth was that the insurance industry [...] had not evolved into a force capable of exerting any serious pressure on the Kyoto process' (Legget 2001: 294). And, notwithstanding a number of encouraging developments, he expresses disappointment that at the beginning of the 21st century 'insurance companies show little sign of being willing to mount a concerted response to the global warming threat' (ibid. 325).

Looking at insurance companies' climate activities some years later, there can be no doubt that things have changed considerably, and that the climate issue has been mainstreamed in corporate strategies. Climate change is largely perceived as a risk, and so is climate regulation. The global consultancy Ernst&Young described climate change as the single most important business risk for the insurance industry (Ernst&Young 2007). Insurance companies in various ways consider climate risks for their business activities and develop strategies to address them, in the first place through a variety of new insurance products that target different dimensions of climate change.

It is necessary, however, to distinguish between different sets of activities here: while many insurance companies assess the risks of climate change for their investment positions, the larger part of their activities addresses the adaptation side of climate change, and focuses on developing new insurance products for extreme weather events and other climate change consequences.

UNEP and the finance sector

About the same time as the Greenpeace insurance industry campaign, the United Nations Environmental Programme (UNEP) started working with companies from the financial sector to address environmental and sustainable development matters. This work resulted in the UNEP Finance Initiative (FI), later supplemented by the Sustainable Energy Finance Initiative (SEFI), and continues until today.

While various UN special bodies deliberated on climate change and its effects in the early 1990s, the economic and finance agencies within the UN system gave less attention to the issue and thus left a blank space for participation of the financial sector (Hohnen 1999). Reacting to this situation, UNEP started to get in touch with banks and other members of the finance industry. Engaging with the private sector this way was not a new approach for UNEP: as one of its founding objectives is to align economic growth with environmental protection, cooperation with businesses is a principle concern for the organisation.

In 1991, UNEP and a small number of commercial banks, including Deutsche Bank and HSBC, created the UNEP Financial Institutions Initiative and launched the *UNEP Statement by Financial Institutions on the Environment & Sustainable Development* at the Earth Summit in Rio de Janeiro in 1992 (UNEP 1997): the financial institutions commit to integrate environmental agreements and objectives in their activities.[117]

Merged later with UNEPs Insurance Industry Initiative to form the *UNEP Financial Initiative* (FI), the partnership brings together more than 200 financial institutions, including most of the global Top 10 banks and insurance companies. Beyond the voluntary commitment to environmental integrity and sustainable development, the main objective of the FI is to make environmental concerns a standard part of investment decisions, by raising awareness of the risks and opportunities related to climate change and other environmental issues.

The core rationale here is the assumption that a better understanding of these risks enables businesses to improve their environmental and financial performance simultaneously. To support investors in this respect, UNEP founded the UNEP Sustainable Energy Finance Initiative (SEFI) that focuses on the specific challenges and barriers for renewable energy investment, including the role of policy frameworks and the use of new finance products. The SEFI convenes financiers and project developers to facilitate deal creation, helps to form public-private alliances that share the costs and risks of those deals, and tries to meet investor concerns by informing on risks and best practices with renewable energy investments, in particular in countries that are barely visible on the renewable energy map thus far.

To do justice to the increasing interdependence between climate change regulation and renewable energy investment, the UNEP FI formed a special climate change Working Group. Beyond training financiers on climate related investments, this group communicates the perspective of the financial sector on the transition towards low-carbon and climate resilient economies to the climate policy arena and, in particular, the UNFCCC negotiations.[118] It focuses on ways to change investor behaviour, emphasizing that financial institutions 'will only systematically integrate climate change factors into core business activities [...]

---

[117] A similar statement for the *Insurance Industry* was launched in 1995 (UNEP 1995).

[118] Towards this objective, it organizes various investor groups that were constituted to speak out on climate issues, and publishes studies on regulation for mobilising private investment for mitigation and adaptation.

154

if these factors are considered to be financially relevant, in other words material'.[119]

Through an *international policy dialogue*, the Working Group wants to 'sharpen the financial materiality of climate change' by supporting both an ambitious international agreement and domestic policies that create pressure to reduce emissions and, thus, require financial institutions to put the issue on their agenda: 'In order for market-based solutions to thrive government has to provide financial institutions with the necessary regulatory architecture. The most important thing governments can do is provide long-term political certainty on regulations, to show the financial services industry that climate change warrants the commitment of valuable time and resources'.[120]

The investment community speaking out

To give more authority to its claims, the UNEP FI delivers its statements to the climate change negotiations together with several investor networks. It aligned with the Europe-focused Institutional Investors Group on Climate Change (IIGCC) and the North American Investor Network on Climate Risks (INCR) to jointly urge governments for an ambitious climate deal in the run up to the climate negotiations in Copenhagen 2009 and Cancún 2010.

Supported by 191 investment institutions that collectively hold 13 trillion US dollars, the 2009 *Investor Statement on the Urgent Need for a Global Agreement on Climate Change* explains that Institutional investors are concerned with climate change and climate policy due to their impacts on both the global economy and individual assets, but are also interested in the opportunities created by the need to respond to climate change: 'Private capital is essential to achieving the transformation to a low-carbon economy [...]. It is therefore critical that heads of state and policymakers understand how climate change-related public policy will influence investment decisions' (IIGCC et al. 2009: 1).

Most importantly, the investor groups emphasize, is to create clear, credible long-term policies that make investors integrate climate change into their investment decisions. This, according to the statement, is thus far hampered by uncertainties about the future of climate change regulation: 'We therefore call on world leaders to reach a strong post-2012 climate change agreement in Copenhagen in December' (ibid.).

---

[119] See UNEP FI website: http://www.unepfi.org/work_streams/climate_change/working_group/index.html.
[120] See UNEP FI website: http://www.unepfi.org/work_streams/climate_change/policy/index.html.

This agreement should include a global target for emissions reductions of 50-85% by 2050, underpinned by developed country targets of 80-95% and lower targets for at least high emitting developing countries to enable and encourage low-carbon investment;[121] reduction targets should also be backed up by national action plans to enhance confidence in investment frameworks. Beyond cost-effectiveness as a general principle and the extension of the global carbon market, the investor groups call for Public Financing Mechanisms that leverage private sector investment, in particular, in developing countries.

One year later, the same investor groups launched another statement concerned with *Reducing Risks, Seizing Opportunities & Closing the Climate Investment Gap* (IIGCC et al. 2010). Whereas the number of signing institutions increased from 191 to 268, the message widely remained the same. Most remarkable is the reference to the 'climate investment gap' between the level of finance needed globally for the low-carbon transformation and the financial contributions pledged in the Copenhagen Accord: 'Without private sector investment, this climate investment gap will not be closed and these objectives will not be achieved' (1). The investor groups reaffirm their interest and willingness to address the climate challenge, but emphasize the 'fiduciary responsibility' that requires them to achieve optimal risk-adjusted returns, and that due to the lack of strong and stable policy frameworks 'many low-carbon investment opportunities do not currently pass this test' (ibid.).

The will to speak out and communicate the position of the finance industry is met by growing interest on the side of governments and governmental organisations to talk about investment conditions for renewable energies and clean technologies. Nick Robins from HSBC observed that the investor statement to Copenhagen was crucial to make policy makers pay attention and listen to investors' concerns more carefully.[122] According to Virginia Sonntag O'Brien from Ren21, a transnational network for the promotion of renewable energies, the work of UNEP FI and a few not-for-profit organisations was crucial for creating an understanding on 'how important policy is for investment decisions' related to climate change, and to raise the interest of both investors and policy makers to address the issue more explicitly.[123]

Before that, a dedicated investment perspective was widely lacking in climate change governance, and so was an understanding on the effect of climate change regulation on investment decisions related to climate protection. It was

---

[121] While the investors acknowledge that most developing countries do not accept such binding emission targets for equity reasons, they highlight that targets reduce policy uncertainty and drive low-carbon investment.

[122] Interview Nick Robins, HSBC.

[123] Interview Virginia Sonntag-O'Brien, REN 21.

this lack of knowledge in particular on the financial and investment side of renewable energy and clean technology governance that made people like Kirsty Hamilton start researching and communicating the perspective of the finance sector

Hamilton worked with the UNEP FI and later the Renewable Energy Finance Project for Chatham House in London. She suggests that the lack of an investment perspective on climate change is due to the dominating role that incumbent industries played in representing the private sector and its interests, but also to the fact that financiers were hardly engaging in climate debates for many years, though they seemed to have 'a lot to say about policy' and a strong interest in regulation.[124]

However, Hamilton's work and the perspective of the finance sector only gained larger attention in climate politics after 2007. The growing attention on an investment perspective in climate change governance can be explained to a large degree on the importance of the low-carbon and climate finance discourses.

When Nicholas Stern and McKinsey gathered a group of experts from governments, IGOs and finance institutions in the run up to the COP 15 in Copenhagen in 2009 to consider the climate finance issue, they quickly agreed that the private side of climate finance was least understood, but an issue that would play a crucial role in the Copenhagen negotiations.[125] Like Hamilton, UNEP FI and a few others, they thus published a study on instruments and regulation to support investment activities related to climate protection (Romani 2009). The proliferation of such studies largely contributed to make an investment perspective on climate protection visible, and to create a particular knowledge of the climate challenge.

4.3.2 Understanding investor behaviour

Together with this knowledge and visibility, the climate investment discourse creates new *forms of identities* or agency positions as well: Increasingly, the so-called investment community appears as a new actor in climate change governance, and with it a whole range of experts that speak on behalf of investor groups and communicate their interests and concerns. The following looks at this specific perspective on the climate challenge, and the particular role of climate change regulation that follows from this perspective.

---

[124] Interview Kirsty Hamilton, Chatham House.
[125] Interview Mattia Romani, Mc Kinsey.

Stimulating investment: it's the policy, stupid!

For Hamilton, the new emphasis on investment is largely due to the growing focus on implementing climate change regulation on the ground. Reducing emissions in the form of low-carbon growth and development, and the deployment of new technologies and energy forms, require driving investment to new industries and sectors.

She criticises that the challenge is often described as 'finding a large pot of money quickly' (Hamilton 2009: 2), as in general, there would be no lack of interest from financiers and investor in climate related investment activities. 'Institutional investors are searching for new asset classes and strategies [...] and the climate economy is emerging as an attractive source of long-term returns' (Robins and Fulton 2009: 145).

However, the challenge is to put in place the conditions that make investors shift their portfolios to clean energy and technologies. Instead of asking for the (global) costs of climate protection, as the Stern Review does, or for the amount of climate finance needed for the low-carbon transformation, an investment perspective on climate protection is a bottom-up perspective that aims at improving domestic investment conditions and frameworks to attract higher shares of global investment capital. Hamilton describes the challenge as creating *investment grade policies* that 'tackle all the relevant factors that financiers assess when looking at a deal' (Hamilton 2009: 1).

Compared to the Kyoto approach to climate protection, an investment perspective on climate change is different in two important ways: on the one hand, it does not aim at achieving global cost-effectiveness in climate protection necessarily, as the objective is to create enabling conditions in different countries simultaneously and independently; and on the other hand, it rejects the expectation that a global price signal like a carbon price will do the job. Whereas it is domestic frameworks and policies that would drive investment and decide on the success of the low-carbon transformation, carbon pricing will only be 'the icing on the cake' (Hamilton 2009: 6).

The problem with relying on carbon pricing from an investment perspective is that it is not expected to change investment behaviour fast enough: 'Even if we get a real strong price signal it takes a while to kick in. Really a lot of the decisions about the technologies we deploy [...] are really driven by a regulatory structure on the ground. [...] The international global climate debate really isn't well positioned to deal with on the ground utility powers'.[126]

---

[126] Bracken Hendricks, Center for American Progress, at the conference *The great transformation – Greening the Economy*, at Heinrich Böll Foundation Berlin, Friday, May 28th 2010 (own documentation).

Governing investment for climate protection is, in a certain way, about guiding market development, as the low-carbon transformation requires the energy industries and other business 'to go in completely uncharged waters'.[127] *Investment grade policies*, in that sense, aim at regulatory frameworks that help create markets for clean technologies, and give incentives for a quick transformation to low-carbon economies.

In some sectors, organising this transformation requires 'policy heavy interventions'.[128] Nicholas Stern therefore calls for an 'industrial revolution that has do be led by policy': National and global policies that provide incentives for clean technology development, price fossil fuels and assist consumers in efficient energy use would 'have the potential to unleash a significant pool of investment that can serve as a powerful engine for a new era of economic growth' (cf. HSBC 2010: 1).

As we have seen above, the awareness of the risks and opportunities of climate change and its regulation enhances the intererst of investors in the climate issue; whereas expectations for an increasing carbon price raise concern mainly with respect to the returns on existing investment position, it is increasingly also the financial opportunities of low-carbon investment that are responsible for this interest (Bales and Duke 2009).

But while it is increasingly claimed that climate change considerations become a standard part of investment decisions, Nick Robins from HSBC observed that the general attention to the importance of climate change and its regulation for investment portfolios is not always met with a proper understanding of the concrete risks and opportunities for investment decisions.[129]

Beyond individual investment positions, the limited investment in renewable energies and low-carbon technologies thus far is also described as a more systematic problem. In a contribution to a book published at Harvard University, Nick Robins and Mark Fulton from Deutsche Bank's Climate Change and Investment Research and Strategy Unit argue that within the investment community, climate change is increasingly seen 'as another example of systemic risk failure on capital markets, with the failure to adequately price carbon being compounded by incentive-driven short-termism. This continues to result in the misallocation of assets to carbon intensive options. Long-term reforms to governance [...] are a necessary complement to deliver capital markets that are fit for purpose for the coming climate economy' (Robins and Fulton 2009: 146).

---

[127] Tom Burke, Founding Director, E3G, at the conference *The great transformation – Greening the Economy*, at Heinrich Böll Foundation Berlin, Friday, May 28th 2010 (own documentation).
[128] Interview Tom Heller, Climate Policy Initiative.
[129] Interview Nick Robins, HSBC.

The government of investment for climate protection would thus have to change the way that investors evaluate investment to climate related activities. In general, investors evaluate clean energy and technology projects just like any other investment, on the grounds of the risk-reward relation, and 'it is worth saying: risk and reward is not the same as cost and benefit' (Hamilton 2008: 9). Within economics, the use of the risk concept goes back to the development of the first insurance mechanisms; rather than addressing an unspecific danger (like in the everyday use of the word risk), it emphasizes that the probability of possible outcomes in a situation of uncertainty can be calculated (Ewald 1991).

Investors, then, are not looking for risk free environments, but conditions in which 'risks can be understood, anticipated and managed' (Hamilton 2009: 3). The risks to clean investment are manifold. Most of the risks that apply to any type of investment projects play a role here as well, such as changing market conditions or innovation cycles, lower outputs than expected, or cross border investment risks related to exchange rates or legal systems.

What makes low-carbon investment special, however, is that returns often depend on low-carbon regulation: regulation is thus part of the risk and return equation (Hamilton 2008). While regulation is considered as a constraining factor for certain types of investment projects, it is the enabling condition for investment in low-carbon technologies and renewable energies in many cases. 'Because only a minority of such investments are inherently financially viable, government-mandated incentives such as carbon pricing, standards, and direct subsidies/feed-in tariffs would be required to generate greater investments in mitigation. The private sector could respond to incentives that provide a high degree of regulatory certainty into the future [...]' (Brinkman 2009: 135).

Investors describe this situation as political risk to investment, and are concerned about policy as investment factor and risk in particular if returns depend on a single subsidy. They assess government regulation not only with respect to its effect on the current economics of a certain project, but with respect to its credibility as well.

Nicholas Stern and many others therefore emphasize that policy has to be *loud, long and legal*: loud in the sense of strong, to improve returns and make investment commercially more attractive; long to reflect the financing horizon of investment projects; and legal as part of regulatory frameworks that investors trust in. 'Improving regulatory certainty', Robins and Fulton (2009: 143) argue, 'is the lowest-cost option' for climate protection and the low-carbon transformation. Keeping things simple is another priority for investors with respect to (climate) regulation. 'The greater the complexity and number of policy variables, the greater the risks that need to be managed' (Hamilton 2009: 6).

Two important distinctions exist regarding the risk perception of investors. First, different types of financiers have different appetites for risk and different expectations for returns; this creates both the opportunity and the need to address investor behaviour more explicitly. Chatham House, the UNEP FI and Bloomberg New Energy Finance, for instance, have jointly published a *Guide for Policymakers* on the investment and risk behaviour of different capital groups like venture capital, private equity and pension funds (Justice 2009).

Second, investment conditions in developing countries are generally described as worse than in developed countries, due to less certain legal frameworks and property rights, and less stable (climate) regulation. However, as some emphasize, investment opportunities in developing countries are often also greater, as major infrastructure investments are needed and absorb large amounts of capital, and faster economic growth in some of these countries allows for higher returns on investments (Brinkman 2009).

In the same sense, the unequal distribution of Clean Development Mechanism (CDM) investment is often ascribed to the fact that those countries that generally receive higher levels of FDI have more developed institutions and investment environments and thus are better prepared to participate in the CDM as well (Boyd, Hultman et al. 2007). In the climate finance debate, a lack of clean energy investment is described as a specific country risk as well that is due to a lack in capacity and conducive investment frameworks (Ward, Fankhauser et al. 2009). Addressing the investment conditions in developing countries thus includes capacity and institution building (Robins and Fulton 2009).

When it comes to supporting private climate finance or investment to developing countries, strategies therefore consider how developing countries can use their own resources to put in place enabling frameworks, following similar programmes in Europe or the US (Robins and Fulton 2009).

These strategies also ask how international institutions can address investment risks in developing countries, 'something like MIGA scaled-up':[130] The Multilateral Investment Guarantee Agency (MIGA), the GEF or the International Finance Corporation as private sector arms of the World Bank offer loans or risk management products and credit guarantees to support cross border investment in developing countries and emerging economies; similar tools are increasingly envisaged for supporting climate related investment as well.

The question that arises with respect to these proposals is how supporting private climate finance flows resembles established practices of Development

---

[130] Interview Ramzi Elias, Project Catalyst.

Cooperation accordingly. Tom Heller from the Climate Policy Initiative suggests that what is required to support the low-carbon transformation in developing countries 'has been done and done badly in a lot of development assistance, so we are not starting fresh. [...] I think what we need is adaptation of past practices'. [131]

4.4 Investment for climate protection: visibilities and invisibilities

The aim of this chapter was to describe and explain the constitution of investment for climate protection as an object of government. It was possible to identify two particular contexts in which governing (private) investment is increasingly seen as crucial for addressing climate change: whereas the climate finance discourse highlights the need to support and scale-up financial flows for climate protection in developing countries, low-carbon strategies aim at driving investment to the low-carbon industries and sectors of the future. A third context, the field of renewable energies and the role it plays in climate change governance increasingly, was crucial for understanding an investor perspective on climate change regulation

To say that investment for climate protection is constituted as an object of government within these contexts is not about neglecting the reality of this object or claiming that the approach chosen to governing investment is wrong or misleading; rather, it is about relating the particular framing of the investment challenge to its contexts of emergence, and to demonstrate that this framing is neither necessary nor arbitrary.

Discourses and practices make their objects visible in a particular way and contribute to a specific knowledge of this object this way; to understand the particular way in which the investment challenge in climate change governance is framed, it is therefore crucial to highlight as well what is obscured or made invisible in the same processes.

In the climate finance discourse, investment for climate protection as an object of government is *made visible* against the need for supporting enhanced climate protection efforts in developing countries, and the understanding that the required levels of finance can be calculated and compared to the existing climate finance sources. The lack of finance that follows from this allows to make private investment a part of climate finance, and to ascribe a crucial role to government in enabling these investment flows.

---

[131] Interview Tom Heller, Climate Policy Initiative.

In the low-carbon growth context, the need for governing investment appears through the objective to accelerate the transformation of economies and the development of new technologies, industries and sectors. As (investment) markets seem not capable of advancing this transformation, more government intervention is required to enable and direct investment to the low-carbon economic processes of the future.

What both contexts make visible are not only the investment *needs* for climate protection or low-carbon growth, but also the investment *opportunities* related to emission reductions. This contributes to making the development of (clean) technologies and industries attractive to a great variety of actors, and consequently the dominant form of addressing climate change.

What becomes more relevant in climate change governance is thus an understanding of ways to govern investment and investor behaviour. The climate investment discourse highlights the importance of domestic frameworks and regulation as enabling conditions for investment. Also, the so-called investment community appears as an actor in climate change governance with specific interests and concerns: This contributes to the formation of a particular knowledge of the investment challenge for climate change regulation.

Whereas investment risks and opportunities are made visible and underpinned with a particular knowledge this way, the low-carbon and climate finance discourses largely obscure, to the contrary, the context in which the government of investment takes place.

The climate finance discourse focuses on the investment and finance flows required for climate protection in developing countries, that is: the lack of a certain type of investment flows, but leaves the patterns of current global finance flows widely unconsidered. Starting from an analysis of existing financial flows and their role for climate protection and the low-carbon transformation could lead to a very different understanding of the financial challenge in climate change governance

A UNFCCC study shows that private sector investment accounts for 86% of current global financial flows, and also that the finance flows required for climate protection in coming years and decades will only be a small fraction of these financial flows.[132] This confirms that private investment must finance most of the transformation to low-carbon economies; it also suggests, however, that the challenge of governing investment for climate protection would be to direct current investment flows towards greater sustainability, instead of focusing on additional (clean) investment needs.

---

[132] A UNEP study likewise estimates that the additional investment required will account for only 3% of global financial flows in 2030 (Ward, Fankhauser et al. 2009).

Such an approach would have to address the current organisation of financial markets as well. Low-carbon strategies describe the low-carbon transformation as an economic opportunity and aim at getting domestic frameworks right for low-carbon investment. Only a few contributions to the GND discourse, to the contrary, point to the financial markets context within which this transformation must take place: The GNDG, in that sense, claims that any Green New Deal or low-carbon strategy must remain incomplete and thus prone to failure without a parallel and radical reform of the global financial system.

Overwhelmingly, however, the approach taken to investment for climate protection is part of, and contributes to, an opportunity turn in climate politics that focuses on the economic opportunities and benefits related climate protection, instead of framing emissions reductions as a constraining factor for economic growth and prosperity; other developments like debates about a green economy point into the same direction.

After describing the techniques and technologies that are deployed in the government of investment (chapter 5), the reflection of the investment dispositif returns to this opportunity turn in climate change governance, and asks in more detail for what is made invisibile in the constitution of investment as an object of government (chapter 6); this serves as the basis for critique in form *of deconstructive genealogy* of the climate investment discourse as well.

5 Governing investment for climate protection

The previous chapter highlighted how investment is constituted as an object of government, through the objective to raise clean investment in both developed and developing countries. This chapter takes a closer look at how various agencies aim at enabling and directing investment for climate protection. The investment turn leads to the creation of new instruments for the government of investment, but also affects existing institutions of climate governance and the approach taken to climate change more in general.

The first section focuses on Public Finance Mechanisms (PFMs) that have been created in recent years with the purpose of scaling-up clean investment flows (5.1). PFMs blend public and private money or use public resources to address the risks and concerns with low-carbon investment: Whereas UK created a Green Investment Bank to improve the domestic investment conditions, the World Bank and other international deploy PFMs to support climate protection and the low-carbon transformation in developing countries.

The second and third sections show that the government of investment for climate protection plays a role in other areas of climate change governance as well: many proposals for reform of the Clean Development Mechanism (CDM) aim at improving the effectiveness and sustainable development contribution of CDM investment through stronger regulation and less market flexibility (5.2); and the design of the REDD(+) mechanism for forest protection faces the double challenge of enabling private investment while maintaining control of the direction and effect of these financial flows (5.3).

It is not claimed here, however, that the government of investment for climate protection follows a coherent strategy and a common interest of all participating agencies; rather, the concrete form of the technologies and techniques of government that address investment is defined within the contexts of their deployment and through the – potentially conflicting – interests and understandings of these actors. Describing this as investment *dispositif* in climate change governance allows understanding the government of investment as lose coupling of a diverse set of practices around an emerging rationality or logic (5.4).

5.1 Public Finance Mechanisms: incentivising private investment

Public Finance Mechanisms (PFMs) take up the objective of stimulating clean or green investment most explicitly, either in the form of joint public-private investment or through different risk sharing and hedging instruments. PFMs play an important role both for driving investment to the low-carbon transformation in developed countries and for supporting low-carbon development in developing countries.

The creation of PFMs for climate protection can build on the experience with similar instruments or *techniques* that target investment flows both within countries and across borders. State funded investment institutions or promotional banks like the German KfW or the Multilateral Investment Guarantee Agency, a member of the World Bank Group, promote Foreign Direct Investment in emerging economies and developing countries through risk insurance and guarantees for investors and lenders. These and similar instruments are increasingly used for financing climate protection as well.

5.1.1 Investment for low-carbon growth: UK Green Investment Bank

Many industrialised countries use policies and instruments to support the deployment of clean technologies and, in particular, renewable energies. Increasingly, the focus is on meeting the investment challenges of these projects. The Loan Guarantee Program that is part of the US Energy Policy Act of 2005 supports the deployment of clean energy technologies through loans and loan guarantees. The Program aims at creating jobs, reducing the dependency on foreign oil, and enhancing economic competitiveness, but does not focus on low-carbon economic activities exclusively.[133]

European countries deploy various instruments to enhance the share of renewable energy. Most prominent are feed-in tariffs (FITs) that guarantee a fixed price for certain types of energy supply or electricity. Rather than direct investment support in the form of capital grants or tax exemptions for renewable energy investments, FITs are seen as an operating support (European Commission 2008). The main addressees of the German FIT, for example, were renewable energy producers, not potential investors; the instrument contributed to the de-

---

[133] The Energy policy act provides incentives for energy from coal and also for oil drillings in the Gulf of Mexico. The focus on enhancing energy security is reflected in the tax reductions as well: nuclear power receives US$4.3 billion, fossil fuel production and renewable electricity production US$2.8 and 2.7 billion respectively.

velopment of a large wind and solar energy industry. Increasingly, however, FIT schemes are seen as a way to attract large-scale investments as well.

The focus is on the Green Investment Bank (GIB) here that is part of the UK low-carbon growth strategy. The GIB was created to support the government in formulating investment policies and frameworks; its primary task is to finance the low-carbon transformation by addressing the market failures and barriers that prevent low-carbon investment. At the time of writing, the GIB was still in the process of formation. The following therefore largely draws on the recommendations of the GIB Commission that was set up for this process.[134]

As with other Public Finance Mechanisms, the task of the GIB is to support, not replace, private investment: 'Wherever private sector activity is viable, the private sector, banks and investors should lead and execute deals. The GIB would act as an enabler for the private sector [and] commit the minimum resources required to support these functions' (GIB 2010: 14). The GIB Commission therefore suggests that the bank should be self-funding and aim for commercial rates of return whenever possible, raising money from capital markets. Additionally, however, the Bank could fund projects on non-commercial terms through grants or loans for, in particular, early stage projects that do not yet attract private funding.

As such GIB operations are split into two different units: the *UK Fund for Green Growth* already supports low-carbon innovation and new clean tech enterprises through grants, subsidies and low-interest loans. The GIB Commission criticises, however, that 'a lack of commercial focus means this investment is not delivering its potential. By channelling most of this funding through commercially structured instruments (rather than non-repayable and poorly focused grants) the UK could achieve double or treble the private sector leverage, while doubling the pace of development. This would allow the UK to [...] open up high growth opportunities to business without significantly increasing the public sector commitment' (26).

The mission of the *Banking Division* is, therefore, more immediately to catalyse private sector investment for the low-carbon transition. To address the market failures and barriers that limit investment to the low-carbon economy thus far, the banking division deploys several finance products:

---

[134] According to an update from December 2011, the British government had committed in the 2011 budget to fund the GIB with £3 billion over the period to 2015. The GIB is meant to evolve over two key phases: government investments managed by a new UK Green Investments (UKGI) team from 2012, starting with waste infrastructure projects and non-domestic energy efficiency projects; and full borrowing powers from 2015.

- The objective of *equity co-investment* in renewable energy projects is to target the financiers of energy infrastructure in the UK that are seen as risk averse. Buying shares in selected low-carbon projects in the development phase, the GIB seeks to lower the concerns of private investors and capital holders, and thus to mobilise additional low-carbon investment. If the investment projects develop successfully, the Bank can sell its shares and reinvest the money to develop more projects.
- Cooperating with private banks to offer *debt* or *loans* to companies or private households (in the case of the latter sometimes complemented with subsidies), GIB funding seeks to overcome investment barriers for energy efficiency projects: the idea is that the main challenge to enhancing energy efficiency is the initial investment, whereas the energy savings made result in quick financial savings.
- So-called *intermediate* or *mezzanine debt* could fill the financing gap for technologies that are proven but yet have no extensive track record, and due to the expected rate of return are not able to raise sufficient equity investment. As the mezzanine debt receives lower returns than normal equity investment, it can raise the returns for other investors and thus leverage additional private sector capital.
- *Risk management* and *insurance* products are expected to unlock larger levels of private investment than direct financial support. One example would be to offer to developers or investors a form of put option that enables them to sell their low-carbon energy or infrastructure assets to the GIB at a fixed price. The financial risk for the government that potentially has to buy these assets above market prices is meant to be met by its influence on the value of these assets, for instance, by introducing a feed-in tariff that fixes the price for renewable energy supply. 'Since these are ultimately Government commitments the Green Investment Bank effectively underwrites these: if it reneges on its commitment, then it will have to take the consequences' (28).

To strengthen its character as a commercial finance institution, the GIB is meant to be independent and not accountable to the government or Parliament for individual investment and lending decisions. But acknowledging that the bank uses public resources and that a certain form of public control may thus be required for legitimacy reasons, the GIB Commissions suggests 'that ministers are allowed to determine overarching priorities', but should not have 'significant influence' over its general corporate policy (37).

The proposed governance structure of the bank is meant to manage this 'tension between investing in the public interest and the need to be commercial' (xv): whereas an advisory council formed of ministers and other public representatives decides on priority sectors for GIB funding, a board of directors alongside finance professionals selects concrete projects or technologies in which to invest.

Altogether, the GIB uses established techniques and instruments for encouraging commercial and investment activities, and also follows market logics as far as possible. Contrary to similar state institutions, however, is that the GIB as the 'world's first investment bank solely dedicated at greening the economy'[135] is focused on specific portfolio of commercial activities only – and seeks to achieve political objectives this way.

5.1.2 PFMs for clean investment in developing countries

We saw that the government of investment has emerged as important task in particular in developing countries. 'A new order of partnership is needed between developed and developing economies', according to UNEP Executive Secretary Achim Steiner, 'one that supports the development needs of developing countries but assists them onto a low carbon trajectory that leap-frogs the 20th century development patterns of the North. [...] Encouraging financial flows between rich and less well off countries is key as is the involvement of the private finance sector and global investment community' (cf. Ward, Fankhauser et al. 2009: 4).

Several initiatives have proposed and developed Public Finance Mechanisms (PFM) to achieve this objective and scale-up financial flows for low-carbon development. Synthesising these proposals makes explicit the particular challenges that the government of investment for climate protection in developing countries has to address. The example of the World Bank Climate Investment Funds, thus far the most important example of a PFM for climate protection, further highlights how these mechanisms are meant to govern investment.

The logic of Public Finance Mechanisms

To understand the general idea behind PFMs, we can draw on the publications from two processes: In the previous chapter we saw that the UNEP Sustainable Energy Finance Initiative (SEFI) brings together investors and project financiers with representatives from governments and international organisations to develop policies that enable sustainable investment, and plays a crucial role in the

---

[135] http://www.bis.gov.uk/greeninvestmentbank.

development of PFMs and guarantee instruments. In the run up to COP 15 in Copenhagen in 2009, the London School of Economics and Nicholas Stern organised a similar public-private discussion to consider the use of public funds for leveraging private investment, gathering representatives from banks and investment companies, consultancy firms, international organisations and the British government.[136]

The discussion led by Stern also brought together many of the people who were involved in the Stern Review process, including representatives from some of the largest finance companies like Deutsche Bank, HSBC, and the investor group P8. Through its members, the outcomes of this discussion fed into the UN Secretary General's High-level Advisory Group on Climate Change Financing as well.[137] As participants anticipated that finance would be one of the core issues on the Copenhagen agenda, they decided to focus on the issue that they identified as least understood: the role of private climate finance, and, in particular, the potential of using public resources to leverage private investment.

Governments, according to this discussion, have two main tasks in meeting the climate challenge (Romani 2009). On the one hand, they have to correct three fundamental market failures: environmental market failures such as the non-pricing of GHG emissions; innovation and technology market failures that 'stem from protecting the private benefits of innovation research' (5); and finance market failures like asymmetric information and high transaction costs that result in a lack of finance for particular types of projects. On the other hand, governments have to meet equity concerns by ensuring 'that developing countries have equal access to private investment' (3).

The challenge that PFMs address, then, is a lack of low-carbon investment in developing countries in particular in the short- and mid-term: Carbon markets are expected to generate the required levels of investment in the long run, but public intervention will be required 'to kick-start investment immediately' (2). Without such intervention, the UNEP SEFI adds, private investment and carbon markets 'are unlikely to meet the needs of many countries and sectors, particularly those in early stages of development' (Maclean, Tan et al. 2008: 12).

PFMs thus follow two different albeit related objectives: To mobilise private investment for low-carbon technology innovation and deployment: the leverage factor on private sector spending is the criterion for success in these

---

[136] The report that summarises the discussion led by Stern was written by private sector representatives together with the UK Government and LSE's Grantham Institute for Climate Change Research (Romani 2009).
[137] Interview Mattia Romani, Mc Kinsey.

170

cases;[138] and to create and scale-up markets for these technologies 'by helping key actors up the experience curve and technologies down the cost curve' (Maclean, Tan et al. 2008: 15): PFMs are expected to mobilise long-term capital flows beyond the investments they directly enable or stimulate.

PFMs are meant to address specific investment challenges only. While carbon taxes, cap-and-trade schemes and other economic instruments would establish the overall economic framework conditions for low-carbon investment, PFMs are rather 'specific barrier removal instruments' (Mostert 2010: 14). The economic justification for PFMs for offering financial support to investors 'is that they reduce the cost to the national economy of achieving a given policy target; the cost of a well-designed public finance intervention is more than offset by savings on the "general framework instruments" side' (ibid.).

The UNEP SEF Alliance calculates that the economic stimulus effect of clean energy and technology investment has a greater effect on economic growth and job creation than investments into traditional economic sectors. Nevertheless, developing clean energy is a long-term task and 'due to legacy subsidies for conventional energy sources, large subsidies for clean energy may be required for many years to offset the embedded subsidies enjoyed by competing energy sources' (UNEP SEF Alliance 2009: v). The public sector will have to make financial commitments for PFMs, but 'substantially less than if it was undertaking the investment itself' (Ward, Fankhauser et al. 2009: 22).[139]

PFMs can take various forms and support low-carbon projects at different stages of the financing chain; several mechanisms can thus complement each other in supporting the development and deployment of a new technology. The UNEP SEFI distinguishes PFMs according to the types of capital that they address (Maclean, Tan et al. 2008):

- *Debt focused* PFMs provide either different credit types or guarantees in those cases where domestic financial institutions, usually banks, have sufficient medium- to long-term liquidity but are unwilling to provide credits due to the risks involved in low-carbon projects.
- *Equity focused* PFMs acquire ownership in a project and, by accepting a subordinated position in profit distribution, can take higher risks than corporate investors.[140] (Public) Equity funds usually make investments in both

---

[138] A UNEP study estimates that PFMs can achieve a leverage effect of 3 to 15, that is, they can mobilise 3 to 15 Euros of private investment for every Euro of public money (Ward, Fankhauser et al. 2009).

[139] Thus far, it is mainly international organisations and national development agencies that run PFMs.

[140] Holders of a *subordinated equity positions* are the first to lose money in case of underperformance and the last to receive returns from successful projects, to raise, if possible, the return of the private

projects and companies, whereas Public Venture Capital funds invest in new technologies at the end of the development phase to open capital bottlenecks in project financing, 'whether this is the early-stage bottleneck globally or the venture-capital bottleneck in the developing world' (33).

* *Grant focused* PFMs target smaller projects mainly, and provide grants for project development that can turn into loans if the project succeeds commercially (an approach that is criticised for promoting a 'lack of business discipline' (34)), or soften project loans through interest subsidies or partial guarantees.

Choosing the appropriate instrument depends not only on the project type, but on the country situation as well (Maclean, Tan et al. 2008): middle income countries often have more developed financial and capital markets with available liquidity and lower cost of borrowing, but risk averse credit practices that slow or hamper the development in new sectors; the task for PFMs is, therefore, in mobilising domestic resources by addressing risks. Least developed countries, at the other end, are often characterised by weak financial markets, highly risk averse financial institutions, and greater country risks due to less stable macroeconomic conditions; the result is a lack in liquidity for financing low-carbon projects and higher borrowing costs, and thus a more immediate need for providing access to capital.

In financing the low-carbon transformation in developing countries, the task of PFMs is also not to duplicate the role of the private sector, but to 'perform[s] roles that the private sector cannot or will not play' (Ward, Fankhauser et al. 2009: 15). The objective is to attract in particular capital from large institutional investors such as pension funds and sovereign wealth funds.[141] 'However, to stimulate their engagement the expected returns on climate change mitigation investment need to be commensurate with the perceived level of risk. This is not currently the case' (ibid. 5).

Different types of low-carbon funds are a particular type of PFM and expected to become an important way of raising capital for the low-carbon transformation in developing countries.

Cornerstone funds are large, commercially managed funds that operate on the regional level and invest in smaller funds like the renewable energy country

---

investors, including institutional investors; this raises concern, however, 'that this model results in removing "too much" risk from the private sector and could blunt the incentives for fund managers and institutional investors to the extent that capital is allocated to projects with little chance of commercial success' (Ward, Fankhauser et al. 2009: 21).

[141] Pension funds control more than US$12 trillion, sovereign wealth funds US$3.75 trillion (Ward, Fankhauser et al. 2009).

funds that finance individual projects. The objective is to attract private capital by reducing investment risks through risk mitigation instruments, or enhancing the returns of private investors through joint public-private investment where the public institution can partially waive its returns (Romani 2009).

Challenge funds offer similar risk and credit enhancement instruments, usually through a Multilateral Development Bank (MDB), but access to these instruments is different: fund managers can bid for support in the form of 'easily accessible and sizeable packages of instruments' such as conditional credits, or guarantees, of first loss equity positions. The MDB chooses those fund managers who offer the largest leverage effect on climate related investment. As challenge funds also aim at the money of large investors, the task of fund managers is thus to aggregate various projects at the fund level, to mitigate investment risks and ensure that sufficient projects are available for funding.

PFMs also address specific risks that investors face in developing countries. Public institutions like the World Bank's Multilateral Investment Guarantee Agency (MIGA) traditionally offer risk-hedging instruments to support exports and investments independently from climate change considerations (Ward, Fankhauser et al. 2009). MIGA addresses the concerns of investors with market environments in emerging economies and developing countries, including government stability and regulatory frameworks, and helps investors 'overcome these challenges in order to achieve attractive and long-term sustainable risk-adjusted returns. Our work is to encourage foreign direct investment [...] by helping investors and lenders mitigate political risks through insurance (guarantees) products' (MIGA 2010: 1).

The World Bank, the MDBs and others enhance this service to support climate relevant investment also (WEF 2009). Framed this way, PFMs can address various country-specific investment risks:

- *Policy risks*, that is, the risk that a supportive regulation like a FIT is withdrawn, can be met through a guarantee or insurance that replaces these payments in the case that the regulatory framework is altered; or by guaranteeing a price floor for 'a key policy variable that crucially affects the profitability of low-carbon investment, for example the carbon price' (Ward, Fankhauser et al. 2009: 17).[142] Such public based guarantees (PBGs) are praised as better suited than other PFMs to deal with investment projects that carry high risks as they do not subsidise these projects, but only jump in if an investment is not successful. And 'at the political-ideological level, supporters of PBGs often point to the government hand-off aspect of these

---

[142] The investor can sell his carbon certificates at this price to the provider of the PFM then irrespective of the current market price.

instruments: the market makes all decisions; politicians do not pick "winners"' (Mostert 2010: 15).

- Currency funds based on public, and complemented with private, money can hedge *currency risks* in countries where this function is not offered by the private sector. Private investment into these funds is more likely if the public shares take greater risk, and a currency fund is effective if it increases the overall leverage effect of public funds on private investment (ibid.).
- A shortage in low-carbon projects is often described as *country-specific risk* as well: 'It is not always (or even often) the case that there is an unwillingness to provide capital for (low-carbon) projects in the developing world per se, but rather that there is a shortage of sufficiently commercially attractive, easily executable deals in which to deploy capital' (19). Often, individual low-carbon projects are too small to attract the interest of investors. Governments could help out in this situation through companies that develop projects in particular in the early stage, similar to the challenge funds described above.[143]

Governments in developing countries are thus expected to play a crucial role in enabling low-carbon investment supported by PFMs, and in determining the direction of financial flows through national action plans that ensure a 'viable deal flow' (Romani 2009: 11). Both national and local institutions can define investment priorities and support investment flows through low-carbon development plans and technical assistance (Maclean, Tan et al. 2008).

Scaling-up low-carbon investment to developing countries is therefore often described as a challenge for developing the institutional capacities and human resources that are required for managing greater investment flows. It is an 'institutional development issue [...]. Lack of domestic sources of capital is rarely the true barrier; inadequate organisational and institutional systems for developing projects and accessing funds are actually the main problem. [...] This entails sustained effort over years' (Taylor, Govindarajalu et al. 2008: 7).

Climate Investment Funds: financing low-carbon development

Several international organisations like the Global Environment Facility and national development agencies launched PFMs to scale-up investment for low-

---

[143] Though these companies should be privately run, as Ward, Fankhauser et al. (2009: 20) emphasise, as the 'skills for successful execution of the role are typically found in private sector companies'.

carbon growth and resilience in developing countries after the Bali COP in 2007, usually in the form of climate funds (Porter, Bird et al. 2008).[144]

The focus is on the World Bank *Climate Investment Funds* here: the World Bank has quickly emerged as the most important player in climate finance and its climate funds also played an important role in changing the approach to finance and investment in climate governance.[145]

The Bank began to expand its climate funding activities based on a mandate given at the G8 meeting in Gleneagles in 2005. In 2008, the Board of Executive directors formally approved the creation of two Climate Investment Funds, the *Clean Technology Fund* and the *Strategic Climate Fund* (SCF). The objective is to enhance the role of the World Bank and the Multilateral Development Banks (MDBs) in addressing climate change, and to integrate climate change concerns in their development funding activities.[146]

While claiming that the MDBs 'can and should play a role in ensuring access of developing countries to adequate financial resources and appropriate technology for climate actions' (World Bank 2008b: 8), the World Bank highlights that its role is in assisting rather than determining mitigation and adaptation activities. The leading role is left to national governments and the UNFCCC process: activities financed by the CIFs should be based on country-led programmes and integrated into country-owned development strategies; also, 'the UN is the appropriate body for broad policy setting on climate change, and the multilateral development banks should not preempt the results of climate change negotiations' (8).[147]

Addressing both public and private sector activities, the CIFs aim at innovative ways of financing 'early transformation climate action' (World Bank 2008a: 6). Whereas the funding principles and the ways of engaging the private sector

---

[144] See for an overview the regularly updated wepage: www.climatefundsupdate.org.

[145] Likewise, the focus will be on financing mitigation, though PFMs are increasingly extended to finance adaptation as well. Within its *Climate Resilient Program*, the World Bank offers credits for making development programs and projects climate proof – though these policies usually are not expected to generate returns, leaving governments with the need to repay the credits from other activities. A study by the Stockholm Environmental Institute found bilateral financial institutions concerned with a 'lack of commercial adaptation projects' (Atteridge, Kehler Siebert et al. 2009: 25).

[146] The funding activities of the SCF are specified in three Programs: The *Pilot Program for Climate Resilience* (PPCR) aims at integrating responses to climate change into development planning; the *Forest Investment Program* (FIP) aims at reducing deforestation and forest degradation; the Program for *Scaling-Up Renewable Energy in Low Income Countries* (SREP) supports low-carbon development and renewable energy use.

[147] Therefore, the SCF is meant to conclude its operation once a new financial architecture under the UNFCCC is effective. However, 'if the outcome of the UNFCCC negotiations so indicates, the Trust Fund Committee [...] may take necessary steps to continue the operations of the SCF, with modifications as appropriate' (18).

are very similar for the different climate funds, the following focuses on the Clean Technology Fund (CTF).

The provision of finance or guarantees to private companies takes place within a country investment plan in which governments, together with an MDB, the private sector and other stakeholder groups, define how CTF finance can help to scale-up low-carbon activities (World Bank 2009d). The investment plans describe the proposed public sector projects for CTF co-financing and the priority sectors and scope of investment for private sector projects. Individual private sector activities can be included after designing the initial investment plan.

The planned public and private sector activities are assessed according to their potential for GHG emissions savings and cost-effectiveness, and investment plans must clearly highlight the transformational potential of the planned activities in describing how policies and regulatory changes address barriers to investment (World Bank 2009e). Investment plans should, therefore: prioritise activities that help scale-up technologies and generate working examples of low-carbon development; explain how the planned investments create policy and regulatory change that will stimulate further low-carbon activities; and estimate the potential for replication in the targeted sectors.

The MDBs can provide finance either to national governments that use the money for policy programmes and reforms or for on-lending to sub-national entities, including the private sector; or directly to sub-national entities. The CTF targets three private sector groups: project developers, including both developers of clean technology projects or companies that integrate those technologies into their operations; investors such as banks, funds and insurance companies; and financial intermediaries that provide credits for clean technology investment. The objective is the same in all three cases: to address the risk-return relation, and to reduce risks or costs that prevent companies from investing or entering a new market.

The fundamental idea behind CTF funding is that early entrants to a new market or technology face higher risks and costs than the later entrants due to a lack of experience with a certain technology or country, or the regulatory barriers that first movers face (they are often, for instance, the first company to negotiate contracts with a regulator). Buying down these costs or mitigating these risks would thus not only enable initial market activities, but demonstrate the viability of a certain technology or activity and thus allow for replication without – or at least with lower – subsidies.

To this end, the CTF offers financing and risk management tools, all of which include a grant element to lower the additional costs of the investment or the risk premium required to make an investment viable (World Bank 2009f). Whereas pure grants are only used to support the elaboration of national investment plans, the main instruments are, first, concessional loans, where the level of concessionality depends on the rate of return that the MDB expects from a project; and second, guarantees that improve investment conditions by mitigating risks that lenders and investors are not willing or able to accept.[148]

The objective, is, however, to intervene into markets or investment decisions only to the degree that is necessary, and thus to avoid distortions and crowd-in the maximum possible level of investment. The CTF does not support projects that are expected to produce losses, therefore; rather, the idea is to support those types of projects where 'the real market risks are lower than the market perceives them to be' (World Bank 2010a: 3), that is, where market participants underestimate the profitability of an investment.

Likewise, the CTF aims at lowering the concessionality of its lending as much as possible and therefore decides on the terms of its lending or guarantee contracts on a case-by-case basis. This can imply to offer different terms to companies within the same country and sector as well if these perceive risks and costs differently; 'the right amount of concessionality is largely a matter of client needs, market conditions and negotiation, and is dependent on information flowing between the companies or being available in the market' (5).

This form of support is applicable not only to project developers, but to investors and finance institutions as well, though the case is slightly different here: whereas a project developer focuses on the profitability of a single project, financiers evaluate this profitability in relation to other possible investments . The CTF funding would thus have to absorb the losses, or better: it would have to compensate for the lower returns that financial institutions expect from a low-carbon investment compared to an alternative use of the same money.

By mid-2011, the MDBs had signed CTF investment plans with 12 governments. Mexico was one of the first countries whose investment plan was approved for CTF funding. The plan outlines a low-carbon strategy and examples for projects that the government seeks to implement together with the World Bank, the Inter-American Development Bank (IADB), and the International Finance Corporation (World Bank 2009b).

The basis for this investment plan are three studies that analyse the economics of mitigation policies in Mexico: a low-carbon growth study funded by Climate Works, the Mexican government, and McKinsey; a World Bank financed low-carbon country case study; and an IADB financed study on the economics of

---

[148] The CTF can provide guarantees for addressing: technical and economic performance risks, for instance in the application of commercially viable technologies in new markets; commercial and financial risks, such as small project scale and weaknesses in domestic capital markets; and country or political risks. However, the CTF does not provide political risk guarantees in public sector projects and rather advocates supporting policy reform, capacity building and technical assistance.

climate change in Mexico. These studies identify priority areas for action, of which the country investment plan selects the transport sector, renewable energy use and energy efficiency.

The investment plan does not focus on new technologies but rather on those that are 'readily available to Mexico today, but face institutional, regulatory, or cost barriers' (9). The plan outlines for the targeted sectors or technologies the emission reduction potential, the regulatory situation, and planned reforms and incentives, and also calculates their cost-effectiveness as the tons of CO_2 reduced per dollar of CTF money invested.

The CTF money addresses the investment challenges of going low-carbon. In the transport sector, for instance, enhancing low-carbon public transport systems could make a big and cost-effective contribution to reducing Mexico's carbon footprint, but the massive initial investments that are needed are not available to the local or regional authorities that often run these systems. The CTF resources could make available this investment capital by blending the loans of the IADB and others with CTF money, lowering the overall interest rate this way.

In the energy sector, one objective is to establish a financing facility in local banks that supports public and private sector investment in renewable energy projects, to demonstrate the commercial viability of these projects. Using concessional CTF funding, once again, is meant to leverage larger IADB loans and money from other infrastructure funding mechanisms, so that the government can provide financial support and guarantees to private companies for developing renewable energy projects, in particular large-scale wind power.

5.1.3 The other side of PFMs: raising public finance

Public Finance Mechanisms are expected to bear costs for governments then, though they provide finance in the form of loans whenever possible. Grants are required, however, to encourage the development of new technologies or for subsidies that make these technologies commercially viable; and costs are expected in particular to finance mitigation and adaptation in developing countries, including regulatory and policy reforms.

The flip side of PFMs is thus the generation of public finance: Many proposals, therefore, ask not only how to spend but also how to raise money for climate protection. We saw that developed countries were rather reluctant thus far to use domestic budgets for financing climate related projects in developing countries. So-called innovative sources of finance attract a lot of interest in the climate finance and investment debate therefore, including proposals for international taxes or auctioning pollution rights (see chapter 4.1.3).

But all creativity in finding new funding sources notwithstanding; governments acknowledge that money from domestic budgets will be required as well. The UK low-carbon strategy, for instance, considers individual savings accounts, levies on energy bills or revenues from emissions trading for financing the low-carbon transformation.[149] The German government, likewise, uses its revenues from auctioning emission permits to finance its national and international climate initiative (BMU 2009).

The challenge to raise money for the low-carbon transformation has also raised interest in the large pots of money from institutional investors, and this has turned the spotlight to the idea of issuing *green bonds*.[150] The rationale is that institutional investors prefer to invest in equities and long-term bonds that offer lower rates of returns but are also less risky, and thus could deliver the required financial resources for climate related projects that often have a long funding horizon as well; issuing green bonds would thus be a way 'to match long term savings with infrastructure investment needs' (GIB 2010: 19).

Green or low-carbon bonds that offer the same level of security as conventional government bonds are thus seen as a promising way of encouraging investors to shift their money to climate protection activities as they can stay with the same asset class. The UK GIB suggests that 'conventional markets have grown up the way they are for good reasons. They are generally liquid and well priced. A green bond market should therefore broadly reflect the existing bond market so that investors can feel immediately comfortable investing in these important assets' (ibid. 20).

The GIB also sees green bonds as a way to prevent domestic investors from buying into bonds issued in another currency due to a lack of adequate domestic bonds: rather than a lack of investment capital, there is a 'real demand for a new type of long-dated instrument' (19).

The World Bank and the European Investment Bank were the first institutions that issued green bonds, both offering a triple-A rating to potential investors. The World Bank launched its first green bond in 2009 within its Strategic Framework for Development and Climate Change, to raise public and private financing for climate related investment projects (World Bank 2009g)'. This was the first time the World Bank offered bonds to raise funds for a specific purpose (Young 2010).

---

[149] By making future increases in tax-free savings exclusive to green *Individual Savings Accounts*: the increasing in saves encouraged this way would be channelled through green investment funds then.
[150] A government that issues bonds borrows money from financial markets at a fixed rate of return.

With this first round of green bonds, the World Bank and the Swedish SEB bank raised US$350 million from Scandinavian institutional investors, including the Swedish National Pension Fund, and various European private banks and life insurance companies. The SEB and the World Bank highlight that the demand for these bonds is driven by a growing interest from institutional and individual investors in climate change related projects in developing countries that fight global warming and poverty. However, the green bonds also offer an attractive risk-reward relation: They guarantee maximum security and an interest rate that, at the time of issuance, was 0.25 percent above Swedish government bond rates.

Also in 2009, the European Investment Bank launched Climate Awareness Bonds, denominated in Swedish Krona, to 'work[s] hand in hand with investors and the banking community' in tackling climate change (EIB 2009). The proceeds from the bonds are used for renewable energy and energy efficiency projects within Sweden and the European Union in the first place. Managed by Swedbank, the Climate Awareness Bonds target Scandinavian investors as well. The annual interest rate is slighly lower than for the World Bank bonds, but still higher than Swedish government bonds. These bonds, according to the CEO of Swedbank, 'offer a unique opportunity for investors to actively make "a green investment" while at the same time enjoying a higher return than through corresponding government bonds with the same high AAA/Aaa rating' (ibid.).

As of mid-2011, the World Bank has issued more than 30 green bonds at a total volume of US$2 billion with various partners. The objective is to create investment products that 'meet investors specific demand' (World Bank 2011): the repayment of the bond investment is not dependent on the outcome of projects, that is, the bond investors do not take over the project risks but earn a guaranteed return.

If governments raise green bonds to finance climate protection or low-carbon projects, the fixed return rates they have to pay to investors mean a 'greater incentive for governments to implement an efficient regulatory framework' (Romani 2009: Summary, 6), as this is likely to enhance the returns of low-carbon projects.

This incentive would be even greater, some suggest, if bonds are issued as indexed linked carbon or climate bonds. This idea was first promoted by the London based consultancy Z/Yen and the investor network London Accord, but also considered in the discussion organised by the LSE and Nicholas Stern (Onstwedder and Mainelli 2010). 'Indexed-linked carbon bonds are emerging as one of the most promising instruments for raising finance on the capital markets, since they provide for genuine government commitment that directly addresses the primary concern of private sector investors' (Romani 2009: Summary, 13).

These bonds would be linked to an index related to climate protection objectives, such as the achievement of a carbon target or the carbon price, and the investor would receive a higher return if the climate target was missed, for instance, if the carbon price turned out to be lower than expected. The idea behind linking the return on bonds to one of these indicators is to give governments an extra incentive to achieve climate policy objectives through strong and stable regulation, as it is public resources that are at risk if these objectives are not met.

The second objective of linking bonds to a climate index is to address the political risks of climate related investment by increasing investor confidence in regulatory frameworks and policies: indexed bonds would be a 'powerful signal to investors of Government intention to deliver climate change policy'[151] as they provide a hedging instrument for climate related investment. 'As a result, this instrument enables investors to put money into projects or technologies that would pay off in a low-carbon future, despite regulatory risks' (Romani 2009, Summary, 13).

However, the proponents of indexed bonds see several weaknesses of the instrument as well, in particular that governments can hardly control factors such as the carbon target or carbon price and thus might be required to pay high extra returns to investors. It is therefore suggested that carbon indexed bonds be used as a 'niche product' (ibid.) only for financiers of low-carbon projects that share the government interest in a higher carbon price or lower carbon level. 'However, because it is a hedging instrument, bond buyers are likely to be the same agents as financiers in emission reduction projects' (Romani 2009, Summary, 7).

---

[151] Green Alliance: http://www.greenalliance.org.uk/grea1.aspx?id=4478.

5.2 Investment and the carbon market

The book argues that the emergence of the climate investment dispositif affects different areas of climate change governance. The regulation of carbon markets, and recent proposals for a reform of this regulation, are a good example for this. The creation, development and governance of carbon markets are related to the objective of governing investment for climate protection in various ways.

On the one hand, carbon market activities were fundamental in bringing about an understanding of the investment opportunities related to climate protection (5.2.1). On the other hand, the objective of financing climate protection and the low-carbon transformation plays an important role in the governance of carbon market mechanisms and its reform: Proposals for reforming the Clean Development Mechanism (CDM) aim at enhancing its contribution to sustainable investment through stronger regulation and less market flexibility (5.2.2). Likewise, the mixed record of carbon trading schemes in financing transformative change has raised calls for more regulation and clearer targets (5.2.3).

5.2.1 Sensing investment opportunities

There is a broad consensus that the creation of a carbon market was instrumental in highlighting the economic dimension of the climate challenge, and for framing it in a way that makes it understandable for market participants. 'Its biggest success so far has been to send market signals for the price of mitigating carbon emissions. This, in turn, has stimulated innovation and carbon abatement worldwide [...]' (World Bank 2008c: 1).

Carbon market activities were also essential for bringing about an understanding of the investment opportunities related to climate protection as they created 'a terminology that business could understand' (Hamilton 2008: 4). Today, a great variety of companies – from large investment banks to carbon finance start-ups, law firms, auditors and consultancies – are seeking returns from low-carbon investments (Paterson 2010). The creation of emission trading systems and the emergence of the carbon market have thus 'mobilized the world of private capital to work in favor of protecting the environment' (World Bank 2008c: 22).

This development is not embraced uniformly, however: 'The establishment of emissions trading schemes has already created a whole new constituency whose vested interests lie not in climate mitigation but in the perpetuation of the schemes and further changes to increase their extraction of profits from the system' (Clifton 2009: 40). Instead of solving the emissions problem, leading

NASA scientist James Hansen has argued, carbon trading 'gives industries a way to avoid reducing their emissions. The rules are too complex and it creates an entirely new class of lobbyists and fat cats' (cf. ibid. 40).[152]

The Clean Development Mechanism (CDM) likewise played an important role in making the investment case for climate protection. Though not all CDM projects achieved the desired emission reduction and sustainable development objectives, the CDM has become 'a hotbed of activity' in climate regulation (Michaelowa and Michaelowa 2007: 1), and it is often emphasized that the mechanism was 'tremendously successful' in raising the interest of very different constituencies and investors in particular (Newell and Paterson 2010: 133).

As with carbon markets in general, this awareness raising is sometimes seen as the most important contribution of the CDM: 'If we see CDM as a limited instrument for creating awareness and commitment worldwide on the issue of climate change, it may be quite successful. If we see CDM as an instrument for achieving sustainable development, it is unlikely that that goal will be achieved. [...] But CDM may be the interim instrument needed to keep the issue alive and investors engaged by meeting their short-term interests' (Holm Olsen 2007: 4).

The World Bank highlights that the growing engagement of institutional investors and the increasing number of funds 'seeking to provide cash returns to investors' has increased the level of investment available for CDM projects, and that CDM projects leverage substantial levels of additional investment for clean energy and other mitigation activities (World Bank 2008c: 3). The CDM, in this regard, is a 'story of unprecedented success' as it has attracted the interest of the private sector in climate related activities and made an important contribution in the creation of a global carbon market (Streck 2009a: 67).

The new interest of investors and the private sector in the opportunities brought about by carbon market development also triggered a more general interest in the climate issue. While low-carbon and carbon market projects are increasingly seen as investment opportunity, uncertainties regarding climate change regulation are perceived as investment risk. Parts of the corporate sector therefore are 'actively seeking clarity on, and supporting definitive government action on climate' (Hamilton 2008: 4); in the last chapter, we saw that investor groups call for stronger climate regulation, and support the growth of carbon markets.

Again, these efforts are not unanimously embraced as contributions to enhancing the environmental integrity of carbon trading and the success of climate protection: 'A tradable permits regime creates new markets which in turn create rents for participants. There is now a rapidly growing set of vested financial

---

[152] Quoted in: Jonathan Leake: The fool's gold of carbon trading. Sunday Times UK, 30 November 2008.

183

interests with every incentive to lobby for the retention and development of the EU ETS' (Helm 2009: 128). This criticism notwithstanding, carbon market development was fundamental for the raising the interest of business and financial actors in climate change.

5.2.2 Governing carbon market investment

The increasing interest in the investment opportunities related to climate protection is accompanied by a shift of the role of carbon markets as part of climate finance. The creation of carbon markets as part of the Kyoto Protocol was embedded in a dominant market efficiency and flexibility narrative that posited the superiority of markets in identifying cost-efficient emission reduction opportunities (Bäckstrand and Lövbrand 2006); this narrative is increasingly replaced by the emphasis that is given to the carbon market as a source of climate finance and an important tool to leverage private investment (Sierra 2007; Pendleton and Retallack 2009; Stern 2009a; WEF 2011).

The limited contribution of carbon offsets to sustainable development

The CDM as the climate finance part of the Kyoto flexibility mechanisms and the carbon markets pursue two objectives: to enhance investment for sustainable development in developing countries; and to lower the costs of emissions reductions for those industrialised countries that accepted binding emissions reduction targets under the Kyoto Protocol.

This double objective results from the contested history of the instrument. The existence of the CDM goes back to an initiative by developing countries that feared to be excluded from the financial dynamics unleashed by the emissions trading schemes agreed on in the Kyoto Protocol, as only countries with binding emissions targets were meant to participate. The original proposal by Brazil for a north-south finance instrument therefore envisaged an independent fund for sustainable development, technology transfer and adaptation, filled by payments from industrialised countries that did not achieve their emission reduction targets.

However, many industrialised countries opposed any financial obligations or penalties from a climate treaty. Due to the pressure of a group of countries led by the US, the Brazilian proposal was transformed into an investment instrument that should serve the interests of industrialised and developing countries alike (Oberthür and Ott 1999): CDM investment into emission reductions in developing countries projects should help developed countries to fulfil their mitigation obligations and foster sustainable development in the recipient countries.

The mechanism proved very successful in achieving the second objective, at least. After a phase of initial reservation, the CDM increasingly attracted the interest of private investors and evolved into the strongest engine of a growing global carbon market (World Bank 2008c). The first and – historically – primary objective of the CDM, however, was left behind in this success story: a number of studies demonstrate the failure of the mechanism to contribute to sustainable development (Muller 2007; Pearson 2007; Sutter and Parreño 2007).

Two main weaknesses of the CDM are widely acknowledged: many projects fail to comply with the criterion of additionality, that is, to prove that the related emissions reductions were only possible through CDM finance; this requirement is often described as difficult or even impossible to fulfil due to the many assumptions it implies about future economic and technological development (Schneider 2007; World Bank 2009a).[153] A second fundamental concern is the uneven regional distribution of CDM projects: while China, India, Brazil and other emerging economies that generally attract high levels of foreign investment host many CDM activities as well, poorer developing countries, in particular in Africa, hardly participate in the CDM (Holm Olsen 2007).[154]

Many assessments interpret the limited success of the CDM as a trade-off between its two objectives: the focus on cost-effectiveness in reducing emissions sets a barrier to financing those projects that contribute to sustainable development substantially and provide social and environmental benefits at the local level. 'The problem is fundamental and stems from the CDM's structure as a project-based market mechanism in which the search for least-cost carbon credits is the paramount consideration. This sidelines projects like renewables by not rewarding the multiple benefits they provide' (Pearson 2007: 247).

This trade-off was anticipated in early CDM studies and confirmed by later assessments, but it must not be seen as a market failure necessarily: the CDM 'is working perfectly in doing what a market-based mechanism is designed to do: discover and direct funding to projects that will produce the maximum volume of carbon credits for every dollar invested' (Pearson 2007: 249).

It is this same success that raises concerns from a sustainability perspective, however. As certificates from the CDM are valued exclusively for their contribution to reduce emissions, investors disregard the complementary sustainability goals associated with the CDM. 'While rhetorically mandated in the Kyoto Pro-

---

[153] The problem with non-additionality in the CDM is that it lowers the level of emissions reductions agreed on in the Kyoto-Protocol: host countries of CDM projects are not obliged to reduce their emissions, so that emissions reductions through the CDM are a zero sum game at best, as an Annex I country can use the reductions for compliance and has to reduce less emissions domestically.
[154] New climate funding mechanisms like the World Bank CIFs are sometimes described as an attempt to overcome the unequal distribution of CDM finance as well (Newell and Patterson 2010).

tocol, they are not monetised and therefore play a limited role in directing investments' (Holm Olsen 2007: 67).

Projects that aim at contributing to wider environmental and social developments are usually more cost-intensive and therefore discriminated against within the market (Gupta 2008). The same is true with capital intensive energy technology options – energy efficiency investments in particular – that likewise are not popular with investors due to high initial costs paired with low returns (Ellis, Corfee-Morlot et al. 2004; Holm Olsen 2007). What was seen as one of the primary strengths of market mechanism turns out to be the central problem then: the high diversity of projects offers profitable investment opportunities but it gives governments limited control in directing resources to the desired places and projects (Wara 2007).

Equally, there is growing recognition that the CDM rules provoke a fundamental conflict for host country governments that have to assess the sustainable development contribution of CDM projects. Taking this task seriously, governments would have to reject potential investment projects if this contribution cannot be proven – though attracting foreign investment is often described as one of the foremost tasks of governments, in particular in developing countries.

The lack of sustainability in the CDM can thus be interpreted in two ways: It is either ascribed to a lack of willingness among developing country governments to take their role seriously and channel CDM investment into the right places by introducing stringent criteria for approval of CDM projects (Michaelowa and Michaelowa 2007: 4). Vice versa, it is understood as a construction failure of the mechanism itself, as 'host countries and their DNAs may have little to bargain with when defining SD standards, due to the global scope of the CDM and investors' wide choice of location' (Holm Olsen 2007: 66, Humphrey 2004).

It is against this background that proponents of the CDM either suggest to lower the expectations of the mechanism and to concentrate on its contribution to cost-effective mitigation, as the CDM 'has been more effective in reducing mitigation costs than in contributing more broadly to sustainability' (Streck 2009a: 70); or to transform the CDM to a sectoral or policy crediting mechanism that no longer supports sustainable development on project basis, but finances the transformation of entire sectors (Reed, Gutman et al. 2009).

Making the CDM work for sustainable investment

The one thing that proponents and critics of the CDM can agree on is that the rules and regulation of the mechanism would have to be fundamentally altered to enhance its contribution to climate protection and sustainable development. How this reform would look, however, is highly disputed, as the World Bank notes: 'Important concerns have been voiced about CDM on issues of its additionality, its procedural efficiency and ultimately, its sustainability. Some critics of the CDM maintain that its rules are too complex, that they change too often and that the process results in excessively high transaction cost; they ask for relief from the rules. Other critics question whether certain project activities are truly additional, or whether CDM can create perverse incentives; they ask for even more rules' (World Bank 2008c: 4).

Those critics that call for less and simpler rules and methodologies usually address the problem that registering CDM projects requires the development of an extra method for approval in many cases, and that it thus often takes two and more years from the registration to the verification of projects; this contributes to 'slowing down innovation and climate change mitigation' (World Bank 2009a: 15). The uncertainties and costs that accompany this process, it is often claimed, are an impediment in particular to smaller projects with local benefits, like the communal use of renewable energy (Boyd, Hultman et al. 2007).

Increasing emphasis is given, on the other hand, to the reforms that can help direct CDM investment to the desired places and uses. In this sense, the climate finance discourse contributes to a new attention for the primary function of the CDM, to deliver finance for sustainable development.

Drawing on analysis in the Stern Review, the World Bank suggests that after appropriate reform, carbon markets could finance 25% of the activities that are needed to stabilize emissions in developing countries. Carbon markets are said to be a particularly powerful tool for climate finance as they can leverage nine-fold investment in some sectors and can also deliver social and environmental benefits; 'the investments brought in by the carbon market can potentially establish low-carbon sustainable development' (World Bank 2008d: 70).[155]

In the climate finance discourse in general, carbon markets are increasingly seen as having a major potential as a tool for financing emission reductions in developing countries. 'Very high expectations are being placed on offsetting through the carbon markets to deliver finance flows for climate mitigation from the private sector in developed countries to governments in developing countries' (Clifton 2009:19). The UNFCCC estimates that by 2020, offsetting could yield up to US$40.8 billion in finance for developing country mitigation (Pendleton and Retallack 2009).

---

[155] In the period from 2002 to 2008, the World Bank estimates, CDM transactions have catalysed over $100 billion of private underlying capital for low-carbon investments (World Bank 2009a).

However, fundamental reform and additional regulation are deemed necessary both to scale-up carbon finance levels and to improve the sustainable development effects of carbon market or CDM investment, according to a UNFCCC study. 'The carbon market, which is already playing an important role in shifting private investment flows, would have to be significantly expanded to address needs for additional investment and financial flows' (UNFCCC 2007: 2).

Joëlle Chassard, who manages the World Bank's carbon finance unit, suggests that 'with the challenge we are facing, needing to reduce emissions much more significantly, we now think that we need to find ways to scale up these mitigation efforts and the use of market mechanisms to achieve large-scale mitigation and emissions reductions' (Euractiv 2009). Likewise, the UN Secretary General's Advisory Group on Climate Finance claims that carbon markets offer an important opportunity to support new technologies and leverage private investment in developing countries. 'The Advisory Group therefore recommends that the carbon markets are further strengthened and developed, while ensuring environmental integrity' (AGF 2010: 11).

One broadly supported step in this direction is to scale-up carbon finance flows through ambitious emissions caps in developed countries that drive the demand for carbon offsets from developing countries (Project Catalyst 2009b). Nicholas Stern emphasises that one crucial advantage of regulating emissions through a global carbon market compared to carbon taxes is that it automatically drives investment to developing countries, instead of leaving it to governments to decide on the level and form of financial transfers (Stern 2009a).

To achieve the desired outcomes through carbon market investment flows, however, a fundamental reform of rules and regulation is also broadly called for. The current carbon market, the World Bank acknowledges, 'is set up for short-term interests and a project-by-project approach, not large, longer-term investments in energy and infrastructure' (World Bank 2008d: 72). The CDM, in particular, 'needs assistance in creating more ambitious and broader incentives for developing countries emission reductions' (Streck 2009a: 72).

Consequently, a great number of proposals to reform the CDM aim at stronger regulation and higher standards in form of sectoral benchmarks, discount factors or positive lists for certain project classes:

• Discriminating different types of project and certificates aims at increasing the CDM legitimacy in terms of sustainability and geographical equity by enhancing the value of social and environmental benefits and emissions reductions in particular countries (Cozijnsen, Dudek et al. 2007; Stripple 2010);

- the reformulation of the CDM as a sector crediting mechanism would entitle all companies within a sector to financial support if they meet a certain standard for energy efficiency or carbon intensity, and transform particular emission intensive sectors like concrete production this way (Boyd, Hultman et al. 2007);
- intermediary carbon banks, run by international organisation or other public agencies, would finance and execute mitigation projects in developing countries and sell the resulting credits (at a higher price) to the carbon market; if the surplus is invested in additional adaptation and abatement activities, these banks would increase the effect of a given level of carbon finance (Stewart, Kingsbury et al. 2009a);
- the taxation of CDM credits or a discount of their value when selling them to the global carbon market would have a very similar effect (Project Catalyst 2009b).

What the different proposals have in common is that they aim at reducing the flexibility of carbon markets in the north-south context: The approach of the initial CDM is to build on the capability of market forces to identify the most cost-effective abatement opportunities by formulating a general goal for emissions reductions and leaving broad space for investment decisions; the proposed reforms, to the contrary, would demand compliance with a greater number of specific criteria.

The reform proposals, then, can be seen to reflect a growing understanding 'that left to itself, the market will not finance high quality projects' (Pearson 2007: 251). The World Bank also suggests that a reformed CDM 'should focus on catalyzing step changes in emission trends, and on creating incentives for large-scale, transformative investment programs' (World Bank 2008c: 6).

The CDM reform would have to aim at specifying the type of activities that are funded: 'Its point of departure must be the promotion of projects that contribute to sustainable development, such as renewables, with the rules and modalities being designed to deliver this outcome' (Holm Olsen 2007: 66).

Proposals for intermediary carbon banks, sector crediting or discounting CDM credits give a stronger role to government authorities in identifying climate related activities and deciding on the priority of different abatement opportunities, and thus in taking investment decisions. As a result, the CDM would be less (than before) characterised by the interests and decisions of market participants, but more by hierarchical forms of government administration and technical expertise. 'These are all activities that most likely will enhance and strengthen the role of states. [...] If the current CDM takes a bottom-up approach, with private

189

actors making project proposals and proposing new methodologies, a reformed CDM can be characterized by more top-down regulation' (Stripple 2010: 79).

Stripple (ibid. 73) rightly claims 'that the [carbon] market can be understood as a form of governance if it is conceptualized as a possibility to structure the possible field of actions of others'. However, this misses one important point. While this governance quality applies to all forms of markets, carbon markets are specific in one sense: 'They do not exist for the sake of it. They are no ends in themselves, but exist as a means to achieve a specific social purpose – to enable societies to reduce GHG emissions' (Newell and Paterson 2010: 142).

The climate finance discourse and the CDM reform proposals bring to mind this specific function of carbon markets again. While the success of carbon markets has been assessed in terms of its overall growth exclusively during its first years, the finance perspective asks for the capability of government authorities to actively decide on the way that the possible field of action for investors is structured and investment decisions are taken consequently.

In a certain sense, then, the current developments make more explicit the role of the CDM and carbon trading as an instrument for governing investment flows. This aspect will be considered in more detail in the reflection of this book (chapter 6).

5.2.3 Enhancing the effectiveness of carbon trading

Without embarking on a broad discussion of the achievements and shortcomings of emissions trading schemes, it can be argued that carbon trading is subject to similar proposals and efforts for enhancing its effectiveness through new forms of regulation. The main focus is in the European Emissions Union Trading Scheme (EU ETS) here, which is by far the largest carbon trading mechanism to date.

The starting point for many reform proposals, likewise, is the difficulties and failures of the current system to achieve the desired outcomes. One general problem of the EU ETS in the first years of its existence was of political rather than technical nature: the system has been highly ineffective not because the trading architecture was inappropriate, but because vested interests managed to lobby their national governments for a more than generous allocation of emissions rights; this watered down the overall ambition of the system substantially (Bailey and Maresh 2009). The European Commissions thus battled for the right to take over the allocation of emission permits from the third trading period after 2012 to achieve a more ambitious cap within the EU, meeting fierce resistance from both member country governments and lobby organisations (ibid., European Commission 2008).

Some critics see this as a permanent problem that is systematically inscribed into the trading mechanism, as its design is based on unrealistic assumptions and neglects the role of interests and power relations: 'Economic efficiency has been used as an argument favouring the trading of pollution permits. The rhetoric of textbook theory has then been adopted as the grounds for creating new multi-billion dollar carbon markets. The divorce between the assumptions of economic theory and complex reality has been neglected' (Spash 2010: 17).

Others, like Nicholas Stern, caution against rejecting emissions trading altogether due to the problems that have arisen: 'Again, [...] the answer is to deal with the genuine issues and refute bad arguments and not to abandon an attempt to use markets with the great advantage they can bring of lowering costs of action' (Stern 2009a: 193).

Beyond this general controversy on the viability of emissions trading as a means of forcing emission reductions, a more fine-tuned debate addresses the difficulties for market participants and investors to make decisions on the basis of carbon pricing through carbon markets. Two related problems have been described in this respect.

First is the problem to factor future revenues from emissions reductions into current investment decisions due to the uncertainty of future carbon prices.[156] The second and closely related problem for investment in the low-carbon transformation is that the carbon price thus far has been too low to sufficiently incentivise long-term investment in clean energy or infrastructure projects; this is particularly relevant, as a report of the World Economic Forum observes, after carbon price expectations have decreased both through economic recession and slow progress in the international climate negotiations (WEF 2011).

Whereas the same publication acknowledges that companies and investors would welcome a price floor, it cautions that 'government needs to be careful not to undermine the market. Left to its own devices, the market will find the right price for carbon and, in this way, help reduce emissions at the lowest possible cost' (WEF 2011: 12).

Others are less willing to rely on the self-regulation capacity of the carbon market, and thus consider instruments to address the problem of low and uncertain carbon prices. Given the limited possibilities for single countries to influ-

---

[156] The same is said for CDM finance for renewable energy projects as 'most banks [...] do not currently see carbon credits as enhancing a renewables project's appeal and are reluctant to lend against a carbon credit purchase agreement' (Pearson 2007: 250); the limited predictability is thus an obstacle for maximising the leverage potential of carbon finance for low-carbon investments (World Bank 2009a).

ence the carbon price of the entire EU ETS, the UK Green Investment Bank Commission advocates 'to provide a risk-reduction mechanism to projects and companies by underwriting a higher and longer-term carbon price beyond 2020. [...] The GIB could help manage these risks on a project by project basis' (GIB 2010: 27).

One option for such a risk-reduction mechanism are long-term option contracts for carbon emissions in the form of put options issued by governments; these guarantee investors and traders a minimum price for their carbon certificates, and thus 'would be a credible commitment to a carbon price floor' (Romani 2009, Chapter 3: 6). Comparable to indexed bonds, these put options would give governments incentives to create regulation and policies that sustain the carbon price, as it would require substantial public resources if the carbon price is lower than expected. Setting a price floor for carbon certificates, then, is about hedging policy risks once again (Ward, Fankhauser et al. 2009).

In a joint publication, the OECD and the International Energy Agency suggest to combine a price floor with a price cap, to reduce investment risks from either too high or too low-carbon prices, and thus to define a certain corridor for carbon prices. 'Price caps on their own, in the absence of a corresponding price floor, create an asymmetrical price risk. This would marginally improve the investment case for a high-emitting coal plant and making the investment case for low-emitting technologies marginally worse. [...] Conversely, price floors on their own would improve the investment case for low carbon technologies and make the investment case for a high-emitting plant worse' (Blyth, Yang et al. 2007: 17).

While proposals for setting price floors for carbon certificates are still being considered in the case of the EU ETS, the Chinese government has already implemented a national price floor for CDM certificates. This way, the government ensures a minimum revenue for CDM projects that are usually implemented through state authorities, and can 'avoid dumping prices. Overall, the state relies heavily on traditional command-and-control and regulates rather top down. Apparently, the Chinese state has captured the carbon market' (Fuhr and Lederer 2009: 338).

Whereas the Chinese government acts like an intermediary carbon bank to capture the gap between the carbon price on the global market and the costs of emissions reduction in China, proposals for a price floor from carbon offsets have been made with the opposite objective as well: to address the fear that offsets could flood the carbon markets and, by undermining the carbon price, lower the incentives for emissions reductions in developed countries.

This concern has often and repeatedly been raised with respect to CDM and was the reason to restrict the use of CDM credits for compliance within Annex I countries. The argument gains new prominence again in particular for offsets from forestry projects. The concern here is that carbon credits from forestry could undermine overall climate protection efforts as long as no reliable MRV of emission reductions through forestry is in place (Karousakis and Corfee-Morlot 2007). Addressing this problem would require either a limit on the amount of forestry credits in the carbon market, an approach taken for the EU ETS with respect to CDM credits, or to underwrite a minimum price floor for these certificates. The latter would enhance investor confidence and thus drive investment to forest projects (Romani 2009).

Though by no means an exhausting discussion of the debates on carbon trading, this quick overview of some proposals for additional regulation shows that carbon trading is increasingly addressed, as in the case of the CDM, from an investment perspective. These proposals consider how to meet the limited capacity of the EU ETS, and carbon trading more in general, as a basis for long-term investment decisions. In that sense, additional forms of regulation and market intervention are considered to improve the contributions of carbon trading schemes to financing climate protection and the low-carbon transformation.

5.3 Investment and forest protection: the REDD mechanism

Carbon market finance is also expected to play a crucial role as a funding source for a mechanism for forest protection in developing countries: The creation of a mechanism for Reducing Emissions from Deforestation and Degradation (REDD) has become a crucial concern in climate change governance through the UNFCCC and beyond in recent years. In the design of this mechanism, one central question is how to attract private investment while guaranteeing environmental integrity.

5.3.1 From Kyoto to Copenhagen: forest in the climate regime

If you go through the initial articles of the Framework Convention, you see many things that relate to a much broader agenda on sustainable development, on food security, on biodiversity, on the protection of ecosystems, and on the preservation of sustainable economic growth. I think we lost our focus for that part of the agenda a little bit in our shift of attention to the Kyoto Protocol. [...] I believe it is really important that we [...] get back to that much broader sustainable development agenda which really was the inspiration of Rio in 1992.

(Yvo de Boer, UNFCCC Executive Secretary, Forest Day at COP 15, Copenhagen)[157]

---

[157] Cf. http://www.cifor.org/publications/pdf_files/cop/cop15/de-Boer-speech.pdf.

At least to a certain degree, Yvo de Boer's wishes became true through the negotiations at the COP 15 in Copenhagen and the development of climate governance thereafter. Forests and forest protection have become a core issue on the climate agenda, and with them the question of sustainable development. 'The simple truth is', HRH The Prince of Wales explained during a high level panel at the Copenhagen Summit, 'that without a solution to tropical deforestation, there is no solution to climate change'.[158]

It was a long way travelled before the forest issue became almost uniformly accepted and embraced as a challenge that should be addressed within the climate regime, though the idea of using forests for climate protection itself is not new. The first studies on the role of deforestation for rising emissions were published in the 1980s and these ascribed 20-40% of global GHG emissions to the loss in forest cover; forests were thus already seen as 'central to an effective management of climate change' (Bäckstrand and Lövbrand 2006: 57).

In political terms, the 1992 Framework Convention on Climate Change encourages countries to monitor, conserve and enhance carbon sinks, including forests. It also contains provisions for a mechanism for north-south financial transfers for forest protection (Art. 4.2a); institutionalised as Actions Implemented Jointly at the first COP in Berlin in 1995, it allows for financial cooperation on forest and sinks projects.[159]

The next and much more important step towards financing forest protection within the climate regime was taken with the adoption of the Kyoto Protocol in 1997. The CDM generally allows financing forest and sinks projects to meet emissions reduction obligations of Annex I countries. Concerns and objections from different parties, however, limited the use of forestry credits substantially, giving a minor role to forest activities in practice.

As no final agreement was possible on the matter in Kyoto due to the strong opposition of developing countries, the decision on sinks in the CDM was taken in Bonn in 2000 and finally adjusted in Buenos Aires in 2004: while afforestation and reforestation became eligible options under the CDM, other land use related activities like forest conservation and avoiding deforestation were excluded.

Parties also agreed that Annex I countries can only offset one per cent of their emissions reduction obligations annually through forestry projects, and that certificates from forestry measures are issued as temporary Certified Emissions Reductions (tCER) that expire at the end of the respective commitment period, contrary to long-term Certified Emissions Reductions (lCERs) for other projects.

---

[158] Own documentation.

[159] Until 2002, 13% of the projects and 35% of the carbon benefits under this mechanism came from forest and land use projects (Bäckstrand and Lövbrand 2006).

The reasons for these restrictions were on the one hand concerns that using a larger share of forestry credits for Annex I country compliance could reduce overall mitigation efforts, as Annex I obligations had already being negotiated and fixed: using more offsets would have reduced the incentives and pressure for mitigation in these countries. On the other hand, there were – and still are – great concerns regarding the environmental integrity of sink projects, in particular with respect to the measurement and verification of emissions reductions and issues such as leakage and permanency (see below). Many developing countries therefore cautioned against a greater share of forestry measures within the CDM.

The main obstacle for financing forests as part of climate protection in the following years were not the provisions of the UNFCCC or the Kyoto institutions, however, but a restriction within the EU ETS: many European governments shared the concerns on measuring and verifying emission reductions from forestry projects, and thus excluded CDM forestry credits from compliance with the EU ETS. As the European carbon market creates by far the biggest (and in fact the only true commercial) demand for CDM credits, this exclusion reduced the overall demand for forestry credits considerably.[160]

Nevertheless, the 'legitimizing discourse' (Bäckstrand and Lövbrand 2006: 60) of the Kyoto architecture and its flexibility mechanisms plays an important role in the approach to forest protection within the climate regime as well. The CDM embodies the ecological modernisation narrative that flexibility and market solutions lower the overall costs of global climate protection and offer a win-win situation for all participants: while developed countries would have to spend less for achieving their mitigation obligations, developing countries could benefit from investment for sustainable development. 'Forest projects developed in partnership between Northern corporate investors and local communities in the South epitomize this flexible mitigation logic' (ibid.).

Equally important for the adoption of the flexibility mechanisms and the development of a forest finance mechanism, according to Bäckstrand and Lövbrand (2006), is an 'operational discourse' of scientific precision that includes the 'techno scientific notion of carbon control' (62).

This expert driven discourse, fostered not least by the IPCC reports and their influence on climate governance, 'represents forests as carbon pools subject to human management, valued according to their sequestration potential [...]. It also tends to represent local project developers as peripheral, subject to strict government and expert control. [...] From this detached perspective, nature is transformed into a tradable commodity and local people in the South are reduced to homogenous project participants' (63).

---

[160] Additional demand comes from voluntary markets and ODA funds, administered mainly by the World Bank.

This management approach to environmental and forest protection that combines cost-effectiveness with administrative rationality is crucial in the creation of a mechanism for Reducing Emissions from Deforestation and Degradation (REDD) as well.[161] While forestry projects were of minor importance in the CDM, mitigating climate change in the forestry sector has moved to the centre of the climate agenda and is now broadly supported by the overwhelming majority of countries, as the Copenhagen Accord affirms:

'We recognize the crucial role of reducing emission from deforestation and forest degradation and the need to enhance removals of greenhouse gas emission by forests and agree on the need to provide positive incentives to such actions through the immediate establishment of a mechanism including REDD-plus, to enable the mobilization of financial resources from developed countries' (UNFCCC 2009: 2).[162] The call for scaled-up, new and additional, predictable and adequate funding that is part of any UNFCCC declaration addresses, in the case of the Copenhagen Accord, the financing of emissions reductions from deforestation and forest degradation as well.

Activities within and beyond the UNFCCC

The forest finance issue returned to the climate agenda during COP 11 in 2005 when Costa Rica and Papua New Guinea proposed the inclusion of a REDD mechanism for forest protection in the climate regime, tied to a financing mechanism filled with contributions from developed countries (Streck 2009b). Thereafter, governments, international organisations and NGOs made a great number of proposals on the design of a REDD mechanism that differ regarding scope (the activities that are included and their definition) and the forms of financial support from the international community.[163]

---

[161] The initial REDD mechanism: Reducing Emissions from Deforestation and Degradation, was later enhanced to REDD+ in many proposals, including conservation, sustainable management and enhancement of forest carbon stocks. As no decision has been taken on the form of the mechanism to be realised under the UNFCCC at the time of writing, this book uses REDD as a generic term for a forest finance mechanism under the UNFCCC.

[162] Not all Parties to the UNFCCC embraced the non-binding Copenhagen Accord, rejecting both the way it was formulated and its (non-) substance. The REDD issue was hardly a reason for these tensions, however.

[163] What all proposals have in common is that they consider policies and financing solutions within the context of the UNFCCC and the second commitment phase of the Kyoto Protocol, that is, against the assumption of abatement obligations for developed countries in the post-Kyoto phase from 2012 that would create demand for forestry credits (Streck 2009b). The little REDD+ book provides a

The Bali Action Plan, adopted at COP 13 in 2007, mandates Parties to negotiate a financial instrument for incentivising forest related mitigation actions in developing countries. Two years later in Copenhagen, forestry was one of the few issues on which Parties could make progress and find at least limited agreement: whereas a draft decision (FCCC/AWGLCA/2009/L.7/Add.6) requests developing countries to take mitigation actions in the forest sector, the decision on how funding is raised and made available for these activities remained bracketed.

With this agreement REDD was seen as one of the few promising issues on the road to COP 16 in Cancún in 2010, and there seemed to be 'an underlying awareness amongst the delegates in Cancún that failure to reach a decision on REDD would result in an implosion in the UNFCCC process' (Parker 2011). Whereas little progress was made in Cancún on substantial issues such as Annex I emissions reductions and climate finance, the COP 16 outcomes – called Cancún Agreements as well[164] – include guidelines and a working plan for establishing a REDD mechanism, de facto adopting the Copenhagen decision.[165]

Parties agreed on a broad scope for REDD as suggested in the Bali Action Plan, including the conservation of forest carbon stocks, sustainable forest management and enhancement of forest carbon stocks. The decision makes reference to the drivers of deforestation but has, as one observer criticises, little to say on how to address these drivers, and neither considers the role that developed countries play in driving tropical deforestation (Parker 2011).

The Cancún decision suggests a phased approach to REDD, beginning with the development and implementation of national plans and forest monitoring systems, followed by results-based actions that are object to measurement and verification. The main issue that remained unresolved in Cancún was funding; the text contains no reference to financial sources, so that the discussion continues to be focused on the two opposing options of markets and funds (Climate Focus 2011). Nevertheless, interpretations of the progress on REDD were widely positive and the agreement seen as a signal that the international community is committed to providing incentives for forest protection (Chasek 2010).

But it is not only in the UNFCCC arena that forest protection has become a major concern in addressing climate change; a great number of initiatives is

---

good oversight of the different approaches and also considers distributional aspects (Parker, Mitchell et al. 2009, see also Rubio Alvarado and Wertz-Kanounnikoff 2007, Streck 2009b).

[164] The decisions were adopted against the clear objection of Bolivia, so that '[w]hether the Cancún Agreements were in fact validly adopted is, however, the subject of some controversy' (Climate Focus 2011).

[165] Draft decision -/CP.16: Outcome of the work of the AWG LCA, section III.C.

moving ahead outside the UNFCCC as well:[166] in the UN Collaborative Programme on REDD (UN REDD), UNEP and the UN Development Programme (UNDP) and Food and Agriculture Organization (FAO) cooperate to prepare developing countries for the adoption of national REDD strategies through capacity building and technical assistance; the World Bank Forest Carbon Partnership Facility (FCPF) supports readiness activities with the Forest Investment Programme (FIP) that aims at both capacity building and concrete forest mitigation activities through investment in the forestry sector and beyond.

Besides international organisations, national governments have started a number of initiatives for financing forest protection in developing countries: during the Copenhagen summit, Norway, Japan, the United States, the UK, France and Australia pledged US$3.5 billion for forestry projects during the period from 2010 to 2012, 'to make early action on REDD+ a reality'.[167] With the REDD+ Partnership, initiated by France and Norway and launched at the Oslo Climate and Forest Conference in 2010, governments agreed on a framework for the rapid implementation of measures for reducing deforestation; by mid-2011, 73 countries have joined the partnership and donor countries have pledged around US$ 4 billion.[168]

Finally, REDD is supported by many non-government organisations as well: the Forests Now Declaration that supports market based mechanisms for protecting ecosystem services and the inclusion of a REDD mechanism in national and international carbon trading schemes is endorsed by a great number of scientists and NGOs;[169] an overwhelming majority of environmental and development NGOs supports the implementation of a REDD mechanism, though sometimes cautioning against too great expectations and criticising specific features of the proposed schemes; and private sector companies from project developers to investors show great interest in a forest finance mechanism as well (WEF 2009).

Against the contested history of financing forests within the climate regime, this broad and almost unanimous support for REDD requires explanation. Instead of asking for the motivations and interests of individual actors, however,

---

[166] For an overview see Karousakis and Corfee-Morlot 2007; Caravani, Bird et al. 2011; and www.climatefundsupdate.org.

[167] http://www.ecoseed.org/en/general-green-news/copenhagen-conference-2009/copenhagen-leading -stories/ 5618-U-S-joins-$-3-5-billion-scheme-to-fight-deforestation.

[168] The REDD+ Partnership is intended to serve as an interim platform for REDD+ to allow for immediate action on capacity building and technical assistance and coordination of REDD+ initiatives and financial instruments, including through a Voluntary REDD+ Database, see http://reddpluspartnership.org/en/.

[169] See http://en.wikipedia.org/wiki/Forests_Now_Declaration, http://www.forestsnow.org/ endorsers.php.

the rush for REDD is explained here as part of the economisation and financialisation of climate politics. The calculation of comparative cost advantages from forest abatement explains the overwhelming expectations to a REDD mechanism; REDD offers a climate protection strategy that provides benefits for all participating countries and organisations by meeting the climate challenge through finance.

Explaining support

It is not a new understanding of the contribution from deforestation and land use changes to climate change that can account for the great support for REDD, as is sometimes suggested (Caravani, Bird et al. 2011): We already saw that studies have presented similar estimates since the 1980s. And it is neither convincing to argue that REDD brings to attention the multiple benefits of forest protection and conservation, including for sustainable development and local livelihoods: this argument likewise has been discussed with respect to payment for environmental services schemes long before the sudden rush for REDD, together with the risks and dangers of these payments (Landell-Mills 2002).

Rather, financing forestry as a means of emission reductions was put on the agenda of governments and potential investors as part of the new emphasis on financing climate protection quickly and in a cost-effective way.

The Stern Review played a fundamental role in supporting and pushing this argument, and it was another report commissioned by the UK government that extended this perspective (Bond, Grieg-Gran et al. 2009, Rubio Alvarado and Wertz-Kanounnikoff 2007): the Eliasch Review on *Financing Global Forests* explains that with an ambitious REDD mechanism, 'the cost of halving global carbon emissions from 1990 levels could be reduced by up to 50 per cent in 2030 and up to 40 per cent in 2050 if the forest sector is included in a global trading system' (Eliasch 2009: xxi, Lubowski 2008).

Many other organisations, including influential agencies in forest governance like the Center for International Forestry Research (CIFOR), support this argument. REDD offers 'high potential at low costs' (Dutschke and Wertz-Kanounnikoff 2008: 1), and 'these lower cost could allow the international community to meet a more ambitious global stabilisation target' (Eliasch 2009: xxi). Consequently, REDD is seen as a 'bridge strategy' that allows to postpone stringent emissions reductions in other sectors and world regions (Lubowski 2008: 1, Karousakis and Corfee-Morlot 2007).

The flip side of the cost-effectiveness argument is that the expected investment for forest protection makes REDD attractive for many developing countries. It was developing countries that brought a forest finance mechanism back on the agenda of international climate negotiations, in particular those countries that would benefit financially from such a mechanism.

Looking at the low-carbon development strategy of a country like Guyana (see 4.2.3) suggests that financing forest protection is increasingly seen as an economic and development opportunity in developing countries as well, instead of a threat to global mitigation efforts. The financial support for forestry projects offers a twofold opportunity for the government of Guyana: REDD payments could be used to develop and improve economic activities in the forestry sector directly, enhancing the production of high-value wood products for export markets and creating opportunities for forest-dependent and indigenous communities; and these payments allow to finance additional low-carbon activities in other sectors, and ultimately to enhance private investment to these sectors as well.

What is puzzling, nevertheless, is the large gap between the huge expectations of the contribution of forests to global climate protection and the many uncertainties and contentious issues that accompany the implementation of a REDD mechanism: whereas the climate negotiations and governance processes beyond the UNFCCC address issues such as temporal and spatial leakage and ask how to enable carbon market and private sector flows to forest protection, NGOs raise more fundamental environmental and social concerns with the form and scale in which REDD is meant to be implemented.

To explain that after years and decades of 'tedious development [...] to advance on the international governance of tropical forests' (Rubio Alvarado and Wertz-Kanounnikoff 2007: 22), a REDD mechanism seems to be able to accommodate so many unresolved questions and conflicting issues, we have to understand the REDD mechanism as actively governing these challenges and difficulties; this includes the creation and maintenance of the investment opportunities related to forest protection.

5.3.3 Governing investment for forest protection

The rush for REDD thus fits neatly into the investment turn in climate governance. Combating deforestation can be understood as an investment challenge as well, as the member of the public-private discussion chaired by Nicholas Stern suggest: 'The root causes of deforestation are economic in nature, stemming from under-investment in the sustainable production of land-based commodities, including agricultural products and timber. Long term investment is required to effect economic transformations that put forest nations onto a sustainable footing' (Romani 2009: 17).

Deforestation, according to the Informal Working Group on Interim Financing for REDD+, is the consequence of a market failure that allows for economic benefits from cutting down trees: 'Correcting this market failure is the key to starting to address deforestation. It will take financial resources on a systemic, international scale to create the right economic incentives for governments, businesses and individuals in developing forest countries to protect standing forests, grow new ones where appropriate, and reduce emissions from deforestation and forest degradation' (IWG IFR 2009: 7).

National REDD strategies build on the same rationale. The government of Guyana calculated that the *Economic Value to the World* provided by Guyana's forests can make a US$40 billion contribution to the global economy annually. As no markets exist for trading the environmental services offered by forests, however, there are strong incentives for individuals and companies to cut down these forests. 'Reconciling this tension between protecting rainforests and pursuing economically rational development is the core challenge that must be addressed to make forests worth more alive than dead' (Republic of Guyana - Office of the President 2010: 7).

Addressing deforestation is therefore about making standing forests economically attractive, and that itself is about financing activities that enhance forest cover and replace those activities that lead to deforestation: 'Such targeted financial provisions are needed throughout the economies of forest nations – from subsistence farmers to multinational corporations' (Romani 2009: 17).

Given the huge expectations to the scope of REDD that is triggered by the cost-effectiveness argument, the required level of finance for enabling REDD activities is likewise huge. One primary task for governments and other public agencies is thus to raise private sector and carbon market finance. Similar to other climate related activities, however, private finance for forest protection is not expected to flow immediately and without support. 'Sustainable forestry and agricultural investments can produce attractive returns for private sector investors, but in many countries there is a significant investment gap between project returns and those required by investors mainly because of high risks regarding political stability and land tenure security' (Romani 2009: 17).

The readiness and capacity building activities of many REDD initiatives thus aim, inter alia, at creating the frameworks that allow for private investment in single projects or national forest programmes, and, ultimately, at creating markets for forest protection activities.

The initial proposal of the Coalition of Rainforest Nations suggests a three phase approach: public funding from Annex I countries for capacity and institution building in developing countries, followed by voluntary funding in the early implementation phase, to a later phase including full implementation through compliance-based market mechanisms. Many other REDD proposals also support such phased approaches, and ultimately to link REDD to carbon compliance markets in different ways (Streck 2009b).

Host country governments are expected to play an active role in the creation of enabling environments and markets. This role includes, on the one hand, to link global, national and local interests: 'Host countries should identify investment opportunities and partnerships with international investors. [...] Because many REDD+ projects are land-based, it is essential to ensure that the interests of local communities are aligned with REDD+ projects' (Romani 2009: 17).

Likewise, governments are urged to use public spending strategically, 'to stimulate and complement private investment by helping to provide basic readiness requirements and reinforcing the enabling environment for investment' (Dutschke and Wertz-Kanounnikoff 2008: 5). More concrete, it is 'imperative that public financing be deployed in such a way [...] to minimize the risk for the private sector in investing and to maximize the returns without hampering socioeconomic and other environmental benefits' (Gutman and Cabarle 2010).

Carbon markets are one tool for enabling investment in forest protection, and it is often suggested that a carbon market based scheme is best suited to raise the necessary levels of forest finance by attracting high levels of private investment (Andersen 2008; Boucher 2008). Mandatory markets 'are often preferred because they would assure long-term, continuous, and predictable flows of finance for REDD projects contrary to voluntary funds' (Rubio Alvarado and Wertz-Kanounnikoff 2007: 18).

The implementation of a REDD mechanism in general and the use of carbon finance in particular, however, raise a number of difficulties and challenges. The government of investment for forest protection must address the complexity and uncertainties of forest governance, and the concerns that exist in particular on making forests subject to carbon markets and investment decisions. Many of these concerns are raised not only by NGOs and carbon market critics but also in the same arenas and by the same actors that support a scaled-up REDD mechanism.

From an environmental perspective, a problem that is far from being resolved is temporal or spatial leakage, that is, the danger that emission reductions in one place are replaced through emissions somewhere else, or later in the same place. If forestry credits were used to offset Kyoto (or Post-Kyoto) commitments, leakage would globally lead to rising emissions. These concerns are widespread, as discussions on monitoring and verification systems within and beyond the UNFCCC show. A REDD mechanism would thus not only have to enable investment flows to forest mitigation activities, but also ensure that deforestation activities do not move elsewhere.

On the social side of REDD, many NGOs call for guaranteeing the rights and safety of local communities and indigenous peoples (Martone 2010), a concern that is regularly addressed in the REDD negotiations and implementation as well. The current rush for REDD and the expectations to its scope raise particular concerns. The availability of high levels of funding attracts the interest of many stakeholders, and may have consequences on forest governance beyond or even contrary to what is intended, such as elite capture of financial benefits, the loss of access to land and a lack of voice in the decision-making processes. 'This is because of the likely scale of the systems envisaged, [...] and the strong environmental, private sector and developed country interests to establish REDD mechanisms quickly' (Peskett, Huberman et al. 2008: 8).

One particular concern in that respect is the rollback of decentralised forest management systems: 'With billions of dollars at stake, governments could justify recentralization by portraying themselves as more capable and reliable than local communities at protecting national interest' (Phelps, Webb et al. 2010: 312-13). In a similar vein, a study by FERN points to the danger that 'rushing REDD processes' will undermine the need for time-intensive governance and consultation processes that are part of other programmes, and thus 'will not necessarily address the underlying causes of deforestation' (Leal Riesco and Opoku 2009: 4). The government of investment for forest protection, in that sense, has to address the possible adverse effects of rising financial flows to the forestry as well.

We saw that concerns are particularly strong for making forests subject to (carbon) market decisions. *Market-based* schemes that imply the full fungibility of REDD credits within global carbon markets and delegate project and funding decisions to these markets 'might suffer from greater efficiency-equity trade-offs' (Peskett, Huberman et al. 2008: 7), as 'paying for a carbon service [...] makes it harder to incorporate issues such as biodiversity and poverty considerations' (Scholz and Schmidt 2008: 3).

Several proposals have therefore been made for *market-linked* schemes that would lower this influence (Scholz and Schmidt 2008). The government of Norway and the Climate Action Network suggest using carbon market finance from auctioning emissions allowances at the national or international level to fund REDD, and to distribute this money through a global fund administered by the UNFCCC; other proposals envisage a separate REDD trading scheme and commitments by Annex I countries to purchase a minimum level of REDD credits from one or several countries without using them for compliance. In market-linked systems, then, carbon markets are a source for forest finance, but not the place where forest funding decisions are taken.

5.4 The investment dispositif in climate governance

This chapter showed that financing activities related to climate protection has become a central concern in climate change governance in recent years, as various actors within and beyond the UNFCCC context aim at enhancing private investment for climate protection. It is crucial to see, however, that the financial challenge in climate governance is approached from different sides and with different specific objectives. The governmental activities to raise or enable climate related investment are not linked necessarily:

• Public Finance Mechanisms address the objective to enable and increase private investment for climate protection and the low-carbon transformation most immediately. They deploy various techniques that are established instruments in other – economic – domains or governance arenas, and seek to enable commercial activities related to climate protection this way.
• Proposals for a reform of the Clean Development Mechanisms highlight that carbon markets and their governance are increasingly assessed from a finance perspective as well; they aim at matching the rising volume of carbon finance flows with the initial objectives of climate protection and sustainable development. This would require, as many of these proposals highlight explicitly or implicitly, to reduce (carbon) market flexibility and enhance the responsibility of states in directing carbon finance to the desired ends.
• In the creation of a REDD mechanism for forest protection, finally, one crucial challenge is to align incentives for private investors with the complexities of forest governance. Proposals to use carbon finance for mitigation activities in the forest sector in a way that does not integrate forestry into carbon markets highlight the concerns with making forests subject to market or investment decisions; this, in a certain sense, also reflect the experience with the Clean Development Mechanism.

Taken together, these government activities contribute to the emergence of an investment dispositif in climate governance that develops around the common objective to enable investment to climate related activities. The dispositif perspective highlights the common ground or rationality of these activities, but does not claim that they are determined by this rationality: Rather, is understands the relation between these governmental activities as a loosely coupling around increasingly common understandings of the climate challenge. Table 1 highlights the differences and common ground in the activities that form the climate investment dispositif.

	PFMs	**CDM**	**REDD**
ISSUE	Climate finance and investment	North-south carbon finance	Finance for forest protection
OBJECT OF RESEARCH	UK GIB World Bank CIFs	CDM reform proposals	REDD mechanism design
SPECIFIC OBJECTIVE	Addressing risks, enabling investment	More control in directing carbon finance	Aligning forest protection with investor incentives
COMMON STRATEGY	Directing investment to specific climate protection activities Role of the state in setting investment priorities		

Table 1: Investment dispositif in climate governance

The next chapter reflects the significance of the climate investment dispositif, asking whether and how its emergence changes the rationality of governing climate change. The focus in this analysis will be on the particular approach taken to investment for climate protection, and on the enhanced responsibility that is ascribed to states in managing the climate crisis by creating enabling environments for low-carbon investment.

6 Markets, investment and the state

This book raised two sets of questions: First, how the discussion around the Stern Review re-frames climate change as an economic challenge and takes effect on climate governance; and second, how this economic debate is related to the new emphasis on raising and enabling investment for climate protection, and whether this changes the market-driven approach to climate protection instituted with the Kyoto Protocol.

The analysis in the preceding chapters addressed these questions from different angles. Chapter 3 highlighted that the Stern Review changed the direction of the economic debate around climate change, highlighting the intrinsic relation between climate protection and economic prosperity. The next two chapters focused on how this discussion is taken up and specified along the objective to enable investment for climate protection, describing this as the emergence of the investment dispositif in climate governance.

The following chapter asks for the significance of these developments within the greater picture of climate change governance. It also provides a critique of the climate investment discourse in form of deconstructive genalogy, highlighting blind spots in the way that climate protection is framed as economic and investment challenge. The focus here, it is worth emphasising once more, is on the logics that drive the turn to finance and investment in climate governance, rather than the prospects of individual strategies.

We have to remember that, according to Laclau (1990), it is no easy task to identify the moments of decision in discourse, nor the alternatives that have not been realised, as discourse modifies them as well. In the climate investment discourse, however, many deviations and frictions are still quite visible, as no single and coherent strategy has yet developed. This makes genealogical critique easier, as alternative understandings can be found within the same discursive space; it also means, however, that critique of the climate investment discourse can only be a preliminary one.

From a political perspective, however, it is all the more important to highlight alternative understandings of the financial challenges related to climate change, as the climate investment discourse is a discourse in making and thus open for political intervention and change. Taking on board some of the invisibilities highlighted in the following would allow for a more encompassing assessment of the relation between investment practices and combating climate change.

This chapter therefore proceeds as follows. The first part argues that the climate investment discourse challenges two important features of the market-driven climate regime: the cost-effectiveness principle and the process of making things the same (6.1). The second part highlights that the objective to obtain greater control over climate related investment flows does not limit the role of markets in climate governance, but rather seeks to enable new market activities related to climate protection (6.2). The last part of the chapter describes the return of the state in climate governance as applying the strategies of the competition state to the pursuit of climate protection (6.3).

6.1 From reducing emissions to building low-carbon economies

One central concern of this book was to explain how the importance of the Stern Review is related to the emergence of the climate investment discourse. The hypothesis in this sense was that the Stern Review, or the discussion it represents, raises the awareness for the economic risks and opportunities related to climate change and climate protection, and thus the interest in new strategies that align climate protection with economic objectives. Improving the conditions for low-carbon investment is seen as one important strategy in that regard.

6.1.1 A new political economy of climate change

Looking at some of the recent contributions on the politics and economics of climate change (Giddens 2009; Policy Network 2009; Stern 2009a; Newell and Paterson 2010) confirms that the Stern perspective is widely taken up in current climate change debates. Highlighting the economic opportunities of a low-carbon transformation instead of the costs of emission reductions is seen as a more viable strategy for climate protection; but also requires new forms of state intervention into the economy.

Explaining change: Economic risks, economic opportunities

Chapter 3 demonstrated that economic perspectives and interests played an important role in climate governance from the very beginning. Nevertheless it was argued, the discussion around the Stern Review contributed to framing climate change as economic challenge, by highlighting the intrinsic relation between climate protection and economic prosperity.

The analysis of the investment discourse in chapters 4 and 5 confirmed that the Stern Review had offered at least two important messages that framed much of the climate change debates in the years after its publication in 2006.

First, Stern emphasises the magnitude of both the economic risks and opportunities through climate change, and thus helped to place the topic on the desks of companies and governments alike. Four years after the publication of the Review, Stern admits that he and his team considerably underestimated the magnitude of the climate challenge, and claims that governments and businesses would address climate change much more decisively if they would understand the full extent of climate risks and opportunities (Gordon 2011). Nevertheless, the Stern Review changed the direction of the economic debate around climate change, from a strategic argument for delaying climate regulation on cost grounds to the significance of climate protection for the future of economic growth and prosperity.

Second, Stern suggests that it is possible to address climate change in a way that does not interfere with economic interests. If climate regulation is done properly, the argument goes, markets will shift investments into climate sound energy forms, technologies and modes of production, and climate protection will become a stimulus for a new phase of economic growth and prosperity (Stern 2009a).

This argument became very popular in recent years. Organizing climate protection in a way that enhances economic wellbeing is what Giddens calls *economic convergence* (Giddens 2009). This convergence is the fundamental idea behind most low-carbon strategies, and an important rational for building renewable energy sectors. Many, like the UK based think tank Policy Network, emphasize the need for a positive understanding of climate protection, as 'to realise a low-carbon future requires a strong political narrative of hope and opportunity' (Policy Network 2009: 13).

Replacing the burdensome cost perspective on climate change with a more positive narrative that highlights opportunities and benefits is seen as an important step for enabling the necessary (investment) frameworks. Though it might be difficult to avoid the question of how much the low-carbon transformation will cost, Hamilton (2008: 14) argues that it is crucial to 'bring[ing] forward much more clearly the investment opportunity, and the benefits in short and medium term'.

The difference between the burdensome discussion of climate protection costs and the more positive investment perspective points to the difference between the assumed global benefits of climate protection and the distribution of costs and benefits between countries or stakeholders. The British think tank The Climate Group, in a study led by Tony Blair, suggests that 'despite the fact that it is widely recognised that the cost of cutting emissions is far outweighed by the cost of doing nothing, concerns over [...] the distribution of the costs has led to further delays' (The Climate Group 2009: 1).[170] Bracken Hendricks from the think tank Center for American Progress likewise laments that 'society as a whole is loosing money by investing in an outdated energy infrastructure [...] because we are so focused on the costs, and on who is gonna bear the costs, and how we are gonna raise the costs of pollution'.[171]

The emphasis on the economic opportunities related to climate protection reminds of the *Green New Deal* narrative that gained momentum as a reaction to the global economic and financial crisis. Kick starting investments into enhanced energy efficiency or low-carbon energy and infrastructures is expected to contribute to economic recovery. In his *Blueprint* book, Nicholas Stern suggests that the 'technologies and investment opportunities of low-carbon growth will be the main drivers of sustainable growth in the coming few decades. These investments will play the role of the railways, electricity, the motor car and information technology in earlier periods of economic history' (Stern 2009b: 207).

A new role for markets and the state

The rising awareness for the economic significance of climate change, combined with the limited success of a mitigation strategy centred on carbon markets, also gives rise to a renewed debate on the role of markets in the climate crisis and its solution. The market-failure concept is central to this debate. The Stern Review describes climate change as 'the greatest example of market failure we have ever seen' (Stern 2006: 1). In his *Blueprint* book, Stern once more urges that '[a]t the centre of economic policy must be the recognition that the emission of greenhouse gases is a market failure' (Stern 2009a: 11).

Correcting market failures to make markets work for descired outcomes and reduce undesired externalities is not a new idea, but an established concept within (neoclassical) economics.[172] Climate change, in this understanding, is as

---

[170] The report supervised by Tony Blair therefore argues that this deadlock is based in a misunderstanding: 'Previous economic analysis has shown the global benefits of collective action; what we do for the first time here is show that these benefits accrue to all countries, with costs more than an order of magnitude lower when there is global participation' (ibid.).
[171] Bracken Hendricks, Center for American Progress, in the conference: *The great transformation – Greening the Economy*, at Heinrich Böll Foundation, Berlin, 28 May 2010 (own documentation).
[172] Nevertheless, the concept it highly contested within economics and serves to defend or deligitimise either a free market approach or public policy in its attempts to correct these market

an externality from markets that do not price GHG emissions correctly and thus give market participants no incentives to reduce their emissions. Giving a price to carbon, and thus emitters an incentive to reduce GHG levels, is the core idea behind all market oriented forms of climate regulation, such as carbon taxes or cap and trade systems.

Like other market failures, Stern explains, climate change is essentially a policy failure, as it would be the role of regulation to put a price on GHG emissions. This understanding highlights the opportunity to correct a market (or policy) failure and make markets work for climate protection. 'A market imperfection, if corrected, overcomes inefficiency – policy can make everyone better off. Correcting a huge imperfection such as this carries huge gains' (Stern 2009a: 198).

Debates on the role of markets in climate governance have largely focused on carbon markets until recently. However, the awareness for the magnitude of the climate challenge, and the subsequent need for a low-carbon transformation of the entire economy, have broadened this debate to the role of markets in climate governance more general. Giddens (2009: 5) emphasises that 'market forces have a much bigger role to play in mitigation climate change than simply in the area of emissions trading'.

The transformation to low-carbon economies, however, requires a 'return of planning' (ibid. 8) and new forms of regulation are required that 'allow market forces to become centred upon promoting environmental benefits' (ibid. 92). An 'industrial revolution led by policy', according to Stern (2009a: 7), must 'provide a framework for the entrepreneurship and discovery across the whole of business and society, which can show us how to achieve a cleaner, safer, more sustainable pattern of growth and development'.

The objective to raise investment for climate protection plays central role in the renewed debates on the appropriate role of markets in climate governance. Hamilton (2008: 5) claims that one fundamental insight of an investment perspective on climate change is that the opposition of free-market and command-and-control approaches in climate governance has 'falsely polarised the role of government', suggesting that regulation would undermine market efficiency rather than enable markets to work for climate protection.

An investment perspective on climate change, apparently, could help to overcome this false dichotomy. Presenting the fourth edition of HSBC's business Climate Confidence Monitor in October 2010, Stephen Howard of the *Climate Group* said that it is 'really exciting to see that around the world, people under-

---

failures. In this respect, *market-oriented* necoclassical economists differs from free-marketeers, as the former advocate to correct market failures whereas the latter urge for the abdication of state control over markets wherever possible (Eckersley 2000).

stand that tackling climate change is good for business and the economy. We agree wholeheartedly with the public sentiment that the private sector needs to invest more, but they need political support to be able to do so. We are working with governments to develop smart policies that incentivise business leadership and unlock private sector financing to help solve this issue'. [173]

6.1.2 From markets to finance in climate governance

The climate investment discourse can thus be understood as one important expression of the search for strategies that align climate protection with economic objectives. The objective to scale-up climate related investment reflects both the magnitude of the climate challenge and the opportunity side of going low-carbon.

The new emphasis on finance and investment for climate protection also effects on the rationality of governing climate change. Chapter 5 identified the common ground of the emerging investment dispositif in climate governance as the objective to direct investment to specific climate protection activities. This challenges what has been described as Kyoto approach to climate protection and 'old' rationality in climate governance. At first sight, neither cost-effectiveness as the fundamental principle of the Kyoto Protocol nor carbon markets as a primary tool are given up; however, the climate investment discourse changes the role of both.

Cost-effectiveness or a greater transformation? Yes, please!

We saw that the 1992 Framework Convention on Climate Change introduces cost-effectiveness as a primary principle for climate protection. The Kyoto Protocol institutionalises this principle through three flexibility mechanisms: The creation of a global carbon market helps achieving a certain amount of global emission reductions (agreed on in interstate negotiations) in the places and through the activities that are most cost-effective. The rationale behind this is that it is possible to combat climate change without targeting particular economic activities directly; instead, it can be left to market forces to identify the most economically rational abatement opportunities.

With the growing awareness for the magnitude of the climate challenge, climate change ceases to be seen as an externality to economic processes or a

---

[173] Cf. http://www.theclimategroup.org/our-news/news/2010/10/27/public-see-the-economic-opportu nity-in-addressing-climate-change-says-hsbcs-climate-confidence-monitor-2010/.

problem that can be addressed through changes at the margins of the economy. Rather, combating climate change requires structural transformations at the very heart of economies. Mitigating climate change, consequently, is at the same time a constraining factor and enabling condition for economic prosperity (see chapter 4.2).

The magnitude of climate change and its intrinsic links with economic processes suggest the need to depart from the Kyoto approach by addressing climate relevant activities, industries or sectors beyond lowest-cost emission reductions. The low-carbon growth narrative, in that sense, highlights the need to transform entire economies: low-carbon strategies give priority to transforming the most emission intensive economic activities, rather than focusing on lowest-cost options only.

Likewise, the CDM reform debate highlights that the success of the CDM in identifying lowest-cost emission reductions, instead of addressing more emission-intensive activities, has lowered the overall ambition of global mitigation efforts. Proposals for replacing the original CDM approach with a mechanism that targets entire polluting sectors or rewards specific contributions to mitigation or sustainable development aim at a more substantial transformation of economies.

While these developments suggest an approach to climate protection that goes beyond lowest-cost emission reductions, the cost-effectiveness objective is also maintained. It is fundamental in the studies and strategies that define an economic rational approach to climate protection, like the Mc Kinsey cost curve, and also plays an important role in the climate investment discourse as well. The comparative cost advantages of climate protection in developing countries are an important argument for scaling-up mitigation efforts in these countries. Likewise, the popularity of the REDD mechanism is strongly rooted in the cost-effectiveness of abatement in the forestry sector, and the opportunity to address climate change through the provision of finance.[174]

The climate investment discourse is thus ambivalent with respect to the objective of cost-effectiveness. This reflects the tensions in defining an economically rational approach to climate protection. The Mc Kinsey cost curve is exemplary for this: The study suggests that to keep global warming below the 2 degrees Celsius threshold, 'all regions and sectors would have to capture close to the full potential for abatement that is available to them' (Mc Kinsey 2009: 6). Nevertheless, it focuses on 'the most economically rational abatement opportuni-

---

[174] While financing forest conservation is sometimes seen as a bridge strategy that allows postponing emission reductions elsewhere, others emphasise that forestry abatement must be additional to the emission reductions in other sectors and in particular in developed countries (see chapter 5).

ties' (ibid.) that cost less than €60 per tonne of reduced CO_2 equivalent and can be realised 'without a material impact on the lifestyle of consumers' (ibid. 7).

The *full potential for abatement* is thus constrained by a certain economic rationality. Defining this economic rationality, and thus an economically rational strategy for climate protection, is no longer left to markets, however. Calculating and comparing the costs of abatement opportunities on a global scale, the Mc Kinsey study rationalises the cost-effectiveness principle in climate governance and makes it accessible for government strategies.

The same is true for the climate investment discourse more general. This becomes apparent in the changing role it ascribes to the Kyoto flexibility and market mechanisms. Carbon markets continue to play an important role in climate governance, but this role is increasingly assessed from an investment perspective: 'For carbon markets to have a substantial effect on emissions', Newell and Patterson (2010: 76) suggest, 'incentives must be created to shift the behaviour of large-scale financial actors'.

Against the magnitude of the climate challenge and the need for transformational change, the role of carbon markets in stimulating and financing low-carbon activities in both developing and developed countries is no longer limited to realising the most cost-effective abatement opportunities. Rather, carbon markets become part of a much broader approach for financing the transformation of entire economies, and stronger government intervention would be required to direct carbon market investment to the appropriate ends (see chapter 5).

Put very simply, the focus shifts from the capacity of carbon markets to identify cost-effective emission reductions, to the investment that is generated in these markets. Both the CDM experience and the REDD discussion suggest that the primary role of carbon markets in the north-south context after the finance turn is the generation of resources; their power to identify least cost abatement opportunities, to the contrary, is increasingly seen as a problem rather than part of the solution. Through the climate investment discourse, carbon market mechanisms become tools for governing low-carbon investment.

Unmaking things the same

The climate investment discourse thus also entails a departure from the logic of commodification and marketisation that is at the heart of the Kyoto approach to climate protection. We saw that the commodification of carbon necessarily contributes to 'making things the same' (MacKenzie 2009: 441) in climate governance: the conditions and circumstances of the production of the commodity are not visible in the abstract form of emission certificates, and therefore play no role in the decisions of market participants.

Chapter 5 highlighted that concerns have arisen with respect to the capability of these market mechanisms to direct carbon finance to the most urgent or effective climate protection activities. Carbon trading and offsets systematically discriminate the more expensive options like the transformation of energy systems or industries, and are increasingly seen as an obstacle to the low-carbon transformation therefore (Lohmann 2009). Several proposals for a reform of the CDM thus aim at highlighting and addressing the diversity of climate protection activities, for example through different types of CDM certificates (see chapter 5.2).

The discussion on the REDD mechanisms for forest protection reveals similar concerns. The difference between a market-linked and a market-based approach reflects the question whether carbon reductions from forest activities will be fully commodified and thus priced within markets, or given a fixed priced. The former would contribute more cost-effectiveness in reducing emissions from the forestry sector, but raises concerns with respect to the type of forest activities that are chosen by market participants. Proposals for market-linked schemes therefore aim at limiting the commensurability and commodification of emission certificates from forestry (chapter 5.3).

The climate investment discourse thus challenges the carbon market logic of the Kyoto Protocol, as it highlight the diversity of mitigation activities and challenges related to climate change. The climate finance and low-carbon discourses emphasise the need for financing activities that contribute to a transformational change of economies, instead of focusing on the cheapest emission reductions only. These discourses also highlight the need for a more active role of governments instead of leaving funding decisions to carbon markets. New governmental instruments and technologies like the World Bank Climate Investment Funds or the UK Green Investment Bank address the risks that have prevented certain types of low-carbon investment thus far, and offer 'tailor-made risk reduction mechanisms' to overcome investment barriers (Koch-Weser 2011: 2).

The CDM reform debate points into a similar direction. The objective to reward CDM activities for being located in specific countries or for contributing to local sustainable development highlights the different quality of emission reduction activities, and is a way of setting preferences. Equally, the support for a market-linked REDD mechanism highlights that investment markets alone are not capable of setting the right priorities for forest protection activities. Many REDD proposal suggest to explicitly reward contributions beyond carbon sequestration as well, such as biodiversity conservation.

The common rationale of the mechanisms and proposals that form the emerging investment dispositif in climate governance is that enhanced state intervention is required to successfully govern investment for climate protection. What is different after the finance turn in climate governance is not that climate protection is about the optimal distribution of scarce financial resources for mitigation activities, but how this distribution is organised. The Kyoto approach to climate protection builds on market flexibility to identify least-cost solutions to climate change; more recently, studies and instruments calculate and rationalise the optimal distribution of financial resources, and thus allow for enhanced state responsibility in organising least cost climate protection.

It is against this background that we can identify changes to the economic rationality in climate governance. The Kyoto Protocol with its focus on markets and flexibility has been described as marketisation or economisation of climate governance (Stripple 2010; Meckling 2011). The climate investment discourse also highlight the economic side of climate change, but ascribes different roles to markets and the state in an economically rational strategy for climate protection. If the markets and flexibility approach of the Kyoto Protocol can be understood as a form of advanced liberal government that 'renders economic growth as the entity to be secured from excessive climate protection costs' (Oels 2005: 201), the government of investment suggests a more intimate relationship between climate protection and economic prosperity.

6.2 Climate protection in a financialised world economy

The climate investment discourse thus challenges the old rationality in climate governance, and reflects current struggles with finding the appropritate role for markets and the state in addressing climate change as the biggest market failure 'the world has ever seen' (Nicholas Stern).

In a sense, it contributes to a renewed discussion on the market-state relation that has gained importance in particular with respect to the (re-)regulation of financial markets after the global financial and economic crisis. (Hart and Hann 2009). The crisis fostered a consensus that uncritical appraisal of market freedom contributed to the crisis, and that some form of taming market powers is necessary to prevent similar breakdowns in the future. In his opening remarks to the UN General Assembly in September 2008, Secretary General Ban Ki-Moon suggested that there is a need for 'less uncritical faith in the 'magic' of markets' (cf. Tabb 2011: 211).

Questions were raised in particular to the self-regulation capacity of markets, given that it was the showpiece of neoliberal free market enthusiasm that broke down in the financial crisis. 'How could it be that today's financial markets, the most sophisticated form of the market probably in human history, could run into trouble of such a massive kind, when the most advanced economic theory had demonstrated that unregulated financial markets will be self-correcting?' (Crouch 2011: vii).

However, while the financial crisis has contributed to questioning (neoliberal) free market enthusiasm 'and at least given the word "neoliberalism' negative connotations' (Brand and Sekler 2009: 5), there is little evidence for a more fundamental challenge to the role of markets in the economy. The focus of financial market reforms in the US and Europe is on stabilising a fragile system, rather than radically questioning its organisation and rationality. Deeg and O'Sullivan (2009: 742) thus suggest that '[a]lthough it is tempting to think of this crisis as a critical juncture in the politics of global finance, the new rules it spawns will promote a fundamental transformation in the way the global financial system operates only if they foster a sea change in the way banks and other financial enterprises go about their business'.

The climate investment discourse reflects similar developments and thoughts with respect to the regulation of climate change. On the one hand, it confirms the limited success of a market-cenred approach to climate protection: New forms of carbon market regulation, in particular of the CDM, would limit the role of market forces in deciding over the direction of carbon market finance. Public Finance Mechanisms like the GIB and the CIFs also seek to bridge the gap between market outcomes as usual and low-carbon economic activities.

On the other hand, the climate investment discourse renews the claim that markets must be at the heart of the solution to climate change. Neither the limited contributions of carbon markets to reducing emissions nor the difficulties with matching investment markets with the low-carbon transformation seem to question the role of markets in climate governance per se.

6.2.1 Governing investment, creating markets

Other differences notwithstanding, the common ground of the government activities that form the climate investment dispositif is the objective to direct (private) investment to specific climate protection activities. This objective could give way to very different strategies: from introducing standards or quotes for certain types of economic activities, to taxing those activities that contribute to the climate crisis, to a transformation of the financial system towards more sustainability.

The climate investment discourse, however, describes the financial challenges related to climate protection as *finance market failure* that results in a lack of investment for particular types of climate related activities (Romani 2009). Addressing the risks related to low-carbon investment would help to overcome this market failure and create new investment opportunities related to climate change. This framing, however, makes visible only a particular dimension of the investment challenge related to climate change, and obscures other important aspects of the relation between investment and achieving climate protection.

Climate policy through investor eyes

Chapter 4 highlighted that the new emphasis on investment for climate protection is explained from mainly two developments. First, the magnitude of the climate challenge, coupled with the failure of carbon markets to induce transformational change to economies and industries, leads to a search for new strategies that transform economic activities more immediately. Creating reliable and conducive environments for low-carbon investment is seen as one important strategy in that regard.

Second is the increasing awareness for the economic opportunities related to climate protection and a low-carbon transformation. Focusing on these opportunities is widely held as a promising strategy for achieving climate protection, as it allows building alliances with the potential winners of the low-carbon transformation, and does avoid conflict with those who benefit from the status quo of the high-carbon economy. 'What's more, those winners – financiers – are rather powerful, and can support you as you build the policies which might produce decarbonisation overall' (Newell and Paterson 2010: 10).

The insight that investors have a perspective on climate change that is quite different from many other businesses is the starting point for this approach. The commercial or business sector is comprised of a variety of actors with potentially very different interests. Even though they all pursue profit maximising interests, whether these actors welcome or reject climate regulation (or prefer certain types of regulation over others) depends on their respective commercial activities (Falkner 2010). Accordingly, what is perceived as *the* voice of business shifted over the years, from denying climate change and undermining emission reduction efforts, to spotting business opportunities and advocating climate change regulation (Newell and Paterson 2010).

The changing positions of business actors and the resulting changes to what is seen as an economically rational approach to climate change highlights the role of agency in the emergence of (economic) rationality, and how a specific economic rationality depends on whose interests are deemed most relevant at a certain point in time.

The increasing emphasis on investment for climate protection thus also reflects the growing importance of financial markets in the economy. The Political Economy literature describes this growing importance through the notion of financialisation more recently. Financialisation characterises the development of capitalist economies in the last decades. While, as Epstein, highlights, no agreed upon definition exists, financialisation broadly refers to 'the increasing role of financial motives, financial markets, financial actors and financial institutions in the operation of the domestic and international economies' (Epstein 2005: 3).

The literature on financialisation is still in its infancy. Nevertheless, it is widely agreed that the process of financialisation is related to economic globalisation and the neoliberal reform era, with the liberalisation and internationalisation of capital markets. According to some, developments of and in the financial markets are even the key driving force behind these processes: 'in the current world economy, finance reigns supreme and neo-liberalism and globalization are themselves expressions of finance' (ibid. 5).

Nölke and Perry (2007) suggest that one crucial dimension of financialisation is the rising share of economic gains that is made through financial activities, a phenomenon they describe as 'profit financialization'. In the US, they calculate, the financial sector's share of all profits rose from 10 percent in the early 1980s to 40 percent in 2007. Another important dimension of profit financialisation is that non-financial firms also increasingly derive their gains from financial activities. Related to this, they argue, is a phenomenon they coin control financialization. In financialised economies, firms are increasingly seen as a collection of assets, which makes the maximisation of shareholder value the primary success criterion for the managers of these firms (Nölke and Perry 2007).

Financialisation thus describes the process 'in which the financial sector grew larger as a percentage of economic activity and expanded its influence over the rest of the political economy' (Deeg and O'Sullivan 2009: 732). Financialisation highlights the increasing dominance of capital and financial markets in the economy, and how this changes the functioning of economic systems at the macro and micro levels (Phillips 2006). One widely shared assumption is that with financialisation, financial elites also gained greater and influence in economic organisation and policy making (Palley 2007).

The climate investment discourse, apparently, develops within the context of a financialised (world) economy. Chapter 4 described how new forms of interaction between government agencies and representatives of the finance industry contribute to the emergence of an investment perspective on climate change, and highlight forms of climate regulation that support rather than constrain business activities

The risk notion is at the heart of this investment perspective on climate change. It allows isolating the obstacles that prevent investors from shifting their portfolios to renewable energies or clean technology development, and to present these obstacles as technical challenges that can be met by targeted government interventions. Investment risks include very diverse things, like the level of returns that can be made from investing in the low-carbon transformation (compared to returns from other investments), uncertainty regarding the development and thus profitability of a specific technology, or the political and macroeconomic circumstances in different countries.

The emergence of the climate investment dispositif highlights the importance of the investor perspective on climate change. Carbon market mechanisms like the CDM and the ETS are increasingly assessed with respect to their how they contribute to creating reliable investment conditions; the design of the REDD mechanism for forest protection reflects similar considerations. Most importantly in addressing the risks related to low-carbon investment, however, are Public Finance Mechanisms like the World Bank CIFs and the UK GIB. One of their core functions is to (partially) take over the risks related to low-carbon investment.

This is not an entirely new strategy for government, but rather a common practice in supporting economic activities and market creation in both developed and developing countries. In the north-south context, organisations like the MIGA or national development banks offer support for investment projects in developing countries that private financiers deem as too risky, or as not offering the required risk-reward relation.

Harvey (2005) highlights that risk-taking through public institutions is an essential, though not always explicit, part of Public Private Partnerships as well that are an important tool in supporting private sector activities and the creation of markets; in particular at the local level, public authorities would often take over the majority of investment risks while private businesses obtain the larger shares of returns.

Enabling markets for climate protection

Government activities that address and mitigate investment risks highlight the crucial role that public authorities play in enabling economic activities and the functioning of markets in many cases, and how deeply public and private activities are intertwined in the market place. Assuming the risks of low-carbon economic activites, governments can contribute to creating new markets where none have existed before, and enable commercial activities in cases where the risk-reward relation is not sufficient to meet the expectation of investment markets.

Creating markets for climate protection, however, starts long before the first money flows or the first contracts for public-private cooperations are signed. The investment discourse does not simply describe and address existing investment risks and opportunities, but contributes to their emergence as well. The assessment of investment risks and opportunities creates space for commercial activities and contributes to the creation of markets.

This performativity of the investment discourse is enacted in framing climate protection as investment challenge. 'In speculative enterprises', Tsing writes, 'profit must be imagined before it can be extracted' (Tsing 2004: 84).

The climate investment discourse entails such imagination. The framing of climate protection activities as investment risks and opportunities helps to imagine and thus create new commercial space. The non-availability of 'bankable projects' (Romani 2009) and the need for 'a pipeline of investment-ready projects' (Maclean, Tan et al. 2008: 6) describe the non-existence of markets for profitable low-carbon investment. If this is understood as a country specific investment risk, this framing creates demand and supply for economic activities at the same time.

Likewise, building institutions and regulatory frameworks contributes to the creation of space for economic and for market activities, instead of simply to the regulation of existing markets. This is in particular apparent in the way that deforestation is addressed after the finance turn in climate governance. The huge interest in the creation of a REDD mechanism for forest protection is due to the objective of cost-effectiveness in mitigating climate change. However, many of the low-cost abatement opportunities in the forest sector do not exist under current legal and regulatory frameworks. It requires fundamental government interventions in form of new laws or property regimes to create cost-effective abatement and investment opportunities in the forestry sector.

Explaining deforestation as a market failure that can be corrected through the provision of finance, as the Informal Working Group on Interim Financing for REDD+ does (IWG IFR 2009), or describing the root causes of deforestation as 'economic in nature, stemming from under-investment in the sustainable production of land-based commodities' (Romani 2009: 17), entails the imagination of economic activities.

It is obvious that these are contingent framings of the challenge to reduce emissions from the forestry sector. Through a different lens, the root causes of deforestation could be described as the overexploitation of forests due to a lack or failure of rules at the community level, or as the global overconsumption of forest products. This would enable and require very different solutions to the problem of deforestation.

The focus on the investment needs of combating deforestation thus produces important blind spots in combating deforestation. Capacity building programs, institutional reforms and forestry projects financed with international support are directed at the drivers of deforestation within forest countries—though there is no lack of evidence that many root causes of deforestation can be found outside the forestry sector, and often also outside the forest rich countries. Whereas the main drivers of deforestation are diverse and differ between countries and world regions, it is often demand from industrialised countries, coupled with the strategies of forest-rich developing countries, that support deforestation (Eliasch 2009).

Combating deforestation coherently would therefore require looking at the demand and supply side of forest economies, and would thus have to address a wider range of (economic) activities. 'Deforestation sounds like a unitary activity, which therefore admits of a unitary solution, but such is not the case' (Giddens 2009: 225). Scholz and Schmidt (2008) claim that 'concerted international action beyond the UNFCCC' is required and 'should include revision of international trade policies, introduction of socio-ecological import standards and reform of agro-fuel quotas and other relevant policies in the EU, the US and elsewhere' (3).

The REDD negotiations, to the contrary, are widely silent on the drivers of deforestation from outside the forestry sector, and in particular on the drivers that can be found outside forest-rich countries. Rather, they focus on creating and financing alternative forms of income production to replace deforestation practices within these countries. This is all the more surprising as many REDD proposals and the UNFCCC negotiations also point to the risk of spatial and temporal leakage, that is, the possibility that deforestation takes place elsewhere or is postponed instead of reduced. However, this is seen as a technical and methodological challenge rather than a systematic problem.

Some critics suggest that the focus on addressing deforestation through finance does not only miss out important causes of deforestation, but might itself contribute to the problem by undermining more substantive efforts: 'throwing money at deforestation without confronting its underlying causes is likely to be counterproductive' (Lohmann 2011). The concern here is that quick REDD processes may undermine the need for more time-intensive governance and consultation processes that are part of programmes like the European Union's Forest Law Enforcement, Governance and Trade Programme (FLEGT).

The reason for this, according to a study by FERN, a NGO specialised on forest policies, is 'a very strong political push to ensure that REDD programmes are implemented fast' (Leal Riesco and Opoku 2009: 2). Given this political pressure and the opportunities it creates, governments in host-countries might be inclined to aim for REDD money rather than the complex processes of programmes like FLEGT. The FERN study therefore claims that 'REDD debates need to focus on how to tackle the underlying causes of deforestation and forest degradation rather than the amount of funding needed to keep forest standing without a discussion on what such funding would be spent on and how it would be spent' (ibid. 5).

These concerns with REDD highlight the role that the expectation of political feasibility and economic opportunities play in the climate investment discourse, and the problem related to these expectations. Like the REDD debate, low-carbon strategies are widely silent on those economic activities that contribute to the climate crisis, and instead seek to de-carbonise economies by building competitive alternatives quickly.

The question is, however, whether increasing levels of low-carbon investment and economic activities cannot go along with maintaining those activities that cause the climate crisis and would have to be replaced. Newell and Patterson, in that sense, suggest that 'climate capitalism needs to seep into capitalism-as-usual and transform it, not be subsumed by it or become an irrelevance: an island of carbon finance in a sea of climate-change-accelerating financial flows' (Newell and Paterson 2010: 187). Apparently, there is a tension between the climate investment discourse as a 'political narrative of hope and opportunity' (Policy Network 2009: 13), and climate change reality where high returns are still possible from high-carbon growth and development.

6.2.2 The need for contesting the power of financial markets

The climate investment discourse as a positive story makes visible the economic opportunities related to climate protection. This, however, is only one side of the relation between investment markets and the challenge to combat climate change. The climate investment discourse does not consider how the current organisation of the (global) financial system contributes to or impedes low-carbon investment. As a consequence, it marginalises proposals for changes to the organisation and rationality of the financial system that could contribute to achieving climate protection and a low-carbon transformation.

A proposal from the UK based Green New Deal Group is one of the few interventions that gained wider attention in the climate investment discourse and addresses the constraints from the current organisation of the financial system for enhancing clean investment flows. It claims that reducing the structural power of capital markets is a precondition for directing investment to climate protection. The finance sector 'will have to return to its role as servant, not master, of the global economy' (Green New Deal Group 2008: 23).

Usually, however, the climate investment discourse treats the financial system as bounded and powerful system, and does not question its rules and rationality. This is certainly not due to a lack of awareness for the role that changing these rules could play for climate friendly development and the low-carbon transformation. A UNEP Policy Brief on the Green New Deal, for instance, highlights that the global financial crisis urges states to reform the international financial architecture, and sees this as a 'unique historical opportunity […] to create the basis of a new Green Economy'. The policy paper does not further address these reforms, however, 'because they are being considered under other relevant policy processes, including under the Commission of Experts of the President of the UN General Assembly on Reforms of the International Monetary and Financial System' (UNEP 2009a: 12/13).

In other words, UNEP highlights the intrinsic link between the organisation of the global financial system and the low-carbon transformation, but refrains from concrete reform proposals, as this is the task of another – expert – discussion.

Strategies for the low-carbon transformation thus have to cope with the conditions and constraints that global financial markets impose on organising transformational change. The UK GIB Commission, for example, suggests that financial markets 'have grown up the way they are for good reasons. They are generally liquid and well priced. A green bond market should therefore broadly reflect the existing bond market so that investors can feel immediately comfortable investing in these important assets' (GIB 2010: 20).

Rather then questioning the rules and organisation of the global financial system, the climate investment discourse suggests a series of technical interventions to make climate protection strategies compatible with the interests and expectations of investment markets. That way, it naturalises and legitimises the current organisation of financial markets.

This framing is not surprising, as it follows a long tradition of expert rule in formulating climate protection strategies. We saw that climate science played an important role in making climate challenge visible and managable as the reducton of global emission levels. Early critics already highlighted that this framing distracts from the concrete socio-ecological circumstances that produce climate change, and replaces the responsibility for the overuse of resouces in certain world regions with the need to manage global emission levels.

The technical approach to climate protection found another important expression in the creation of carbon markets. The carbon market approach is built on abstracting from the specific characteristics of the projects or activities that are traded in emission markets. The evolution of the Kyoto flexibility mechanisms also gives important functions to a whole range of experts who set up systems of auditing, measurement and accounting for mitigation activities (Lovell and MacKenzie 2011), or create the financial products that can be traded in carbon markets (Descheneau and Paterson 2011).

The climate investment discourse, in that sense, recruits a new expert group for the formulation of climate protection strategies, to identify, delineate and calculate the risks involved in investing in climate related activities, and to develop policies and instruments that help mitigating these risks. This approach confirms the general understanding of financial activities as highly technical domain.

De Goede (2004, 2005) describes the evolution of finance into one of 'modern society's most depoliticized areas of activity' (de Goede 2005: 2): The increasing complexity of financial practices and its 'reduction to calculability' (ibid.) suggest that these are beyond public comprehension, and consequently cannot be subject to public debate. The government of investment, likewise, appears as a highly technical undertaking that is widely left to financial experts, and not part of wider public debates yet.

Politicising investment for climate protection

The climate investment discourse makes an important contribution to our understanding of climate change, by highlighting the crucial role of investment markets for achieving climate protection and a low-carbon transformation. Thus far, however, it frames the investment challenge related to climate change narrowly as the need to address the interests and concerns of investors with low-carbon activities – and marginalises alternative ways of making financial markets compatible with the climate challenge.

A good starting point for considering these alternatives is the distinction of different types of investment capital. Thus far, this distinction is made with respect to different risk and return expectations, and serves to address the interests and concerns of capital owners through specialised financial instruments. However, one could also consider different capital types with respect to how their use can become subject to public concern and decision making (Lohmann 2009). This would require opening the black box financial markets, to identify and address the many ways in which different actors in society are involved in investment decisions.

One important entry point are the many ways in which governments on the local, regional and federal levels take investment decisions or influence on these, through public investment programmes, state owned banks, or pension funds.

In Germany, for example, publicly owned retail and investment banks account for almost 30 per cent of the banking sector; thus far, however, they are not required to purse sustainable investment practices, and likewise have no mandate for financing urgent climate related transformation processes like the transition of the energy system. Likewise, an enquiry of the Green Party to the German government revealed the enormous financial assets owned or controlled by the federal government, but also the lack of a definition, let alone a strategy, for using this money in sustainable ways (Deutscher Bundestag 2009).

Setting ambitious sustainability standards for all public economic and financial activities could make a fundamental contribution to a more sustainable financial system, and also provide a great number of best practice examples for other financial market actors.

A second important starting point for greening financial markets are the interests of those whose money is invested through banks and investment funds. The climate investment discourse, for instance, gives great emphasis to institutional investors and their specific, usually risk-averse investment behaviour. The owners of these 'big pots of money' are pension funds in many cases that manage the money from large groups of (western) societies. Debates on how to direct investment to climate protection and the low-carbon transformation would thus also have to consider the responsibility of these capital owners, and their willingness to make environmental goals a priority for investment strategies.

There are many other entry points for opening the black box financial markets, and there is also no lack of proposals for environmental standards in the banking and finance sector—however, very few reforms have been implemented yet. It is beyond reach for this book to assess the usefulness and feasibility of different proposals. Nevertheless, it is worth highlighting that these proposals can contribute to an understanding that investment decisions are not the sole responsibility of a few financial experts, and thus to politicising the climate investment discourse. This alone would not suffice for a fundamental transformation of the financial system, but it could be an important first step in contesting the picture of a bounded and powerful financial system whose organisation and rationality are beyond change.

6.3 The competition state in climate governance

The first part of this chapter highlighted that the climate investment discourse challenges the Kyoto approach to climate protection, by giving greater responsibility to states in directing finance to climate protection, and in orchestrating a low-carbon transformation. However, we just saw that the new role of the state is defined and constrained by the context of a financialised world economy: Facing a powerful global financial system, governments need to get domestic investment frameworks right to increase the shares of low-carbon investment.[175]

To reflect the significance of the return of the state in climate governance, we have to come back to the understanding of the state deployed in this book, and to how Foucault addresses the question of the state within the analysis of governmentality. Foucault claims that the state has no essence, no ahistorical nature, but is, in its concrete forms, always the expression of a specific rationality of power. It is therefore not sufficient to highlight that a greater role for state regulation for climate protection is called for, and that the state has apparently returned as a central actor in climate governance. The return of the state is not self-explaining, because *the* state does not exist.

We thus have to ask for the specific understanding of the state that is entailed in its new responsibility in climate governance more recently, in particular in the climate investment discourse, and how this responsibility changes the rationality of governing climate change.

It is in that sense as well that we can draw on more specific conceptions of the state, for instance from Political Economy literature. While the claim that the state has no essence denies a definition of the state on the level of ontology, it is nevertheless possible to work with ontical state concepts, such as the welfare state or the competition state (see below): these define the state within a concrete socio-historical context, and thus within particular rationality of power.

6.3.1 Going low-carbon as new competitiveness

Taking up Oels´ suggestion that 'climate stability was the entity to be secured by biopower, while advanced liberal government renders economic growth as the entity to be secured from excessive climate protection costs' (Oels 2005: 201), we can say that the climate investment discourse suggests a more intimate rela-

---

[175] In the north-south context, international institutions and agencies support developing countries in improving these frameworks This is not exclusive to the climate investment discourse, but increasingly plays a role in more or less related policy fields such as forests, risk reduction and sustainable development as well.

tion between economic and climate protection objectives. Rather than securing economic growth from climate protection costs, the objective is to enable new forms of economic growth and prosperity.

The climate investment discourse matches with more recent understandings of ecological modernization therefore, according to which 'states would seek to enhance the competitiveness of industry by unilaterally increasing rather than decreasing the stringency of environmental regulation' (Eckersley 2004: 69). The government of investment with its emphasis on tailor made investment regulation can be understood as *smart regulation* that aims at the optimal environment for the growth and competitiveness of domestic (gren) industries (Jänicke 2008).

The government on investment describes a global win-win constellation therefore, as both developed and developing countries are expected to benefit from creating the conditions for building low-carbon economies. The green economy discourse also highlights the opportunities to attract foreign investment for the development of green sectors and industries, in particular for developing countries.

What challenges this win-win narrative, however, is the expectation of increasing competition for low-carbon investment. Hamilton embraces this as 'new competitiveness', where attracting clean energy investment becomes a strategic issue for economic, industrial and foreign policy: 'One might say that under a notion of the "old competitiveness", climate policy poses threat to business, while in a period of significant change, the "new competitiveness" focuses on the sectors and solutions that should be prime drivers of both' (Hamilton 2008: 15).

We can characterise and reflect the new role and responsibility of the state in climate governance by introducing a concept from Political Economy literature. The climate investment discourse suggests enhancing the strategies of the *competition state* to the objective of achieving investment for climate protection and the low-carbon transformation.

The notion of the competition state was coined by Cerny, who observed that 'greater structural interpenetration of national economies' forced changes in (economic) policy that in sum led to a transformation from welfare states to competition states (Cerny 1990: 225). State competition for resources is not a new phenomenon as such. However, economic globalisation has enhanced the importance of competition and changed its character, and has led to a 'pervasive belief in national competitiveness as the means for generating economic growth and rising living standards' (Palan and Abbot 1996: 5).

The rising importance of competitiveness strategies is thus ascribed to the neoliberal turn and the subsequent liberalisation of markets and economies as well. Though neoliberalism originally appeared to be a nation-state-level phenomenon primarily, 'its rise also coincided with structurally transformative transnational and globalizing developments', and ultimately turned 'into the political driving force behind globalization itself' (Cerny 2008: 2).

In a way, the competition state is at odds with (neoliberal) free-market doctrines, as creating the conditions for competitive domestic firms and industries requires to limit market freedom to a certain degree; economic policy in the neoliberal era, in that sense, comprises a strong mercantilist element in supporting the domestic economy and protecting it from international competition (Hirsch 1995, Brand 2009b).[176]

It is important to see that the understanding of competitiveness and the competition state has changed more recently, as it is in particular a more recent understanding of the competition state concept that seems to be fruitful for interpreting the government of investment.

The notion of competitiveness comes from business school literature and focused on the competitiveness of the firm (Lall 2001). In the early application of the concept to the state, this focus on the (domestic) firm remained, and the role of the state was assessed with respect to supporting and enabling exports and foreign investment of domestic industries.

This competitiveness as *aggressiveness*, however, is increasingly replaced or complemented by competitiveness as *attractiveness* (Fougner 2006). In the context of globalised financial markets, the challenge for states is not exclusively to promote the competitiveness of domestic industries and firms, but also to provide an attractive environment for foreign companies or investment. This development is intrinsically linked to the growing importance of transnational corporations and investment capital in increasingly global markets.

Competitiveness as attractiveness, then, describes the role of the state in financialised world economy, and focuses on the role of states in coping with the increasing influence and importance of mobile financial capital.[177]

The emergence and development of this latter type of the competition state in particular does not leave the state in its interior untouched, but has important consequences for its configuration. It affects, on the one hand, the orientations of its actors. Cerny and Evans (1999) highlight that social democratic governments deploy the same strategies as conservative governments to handle what they

---

[176] This is complementary to the role of the neoliberal state in creating markets and providing the conditions for their operation; the competition for low-carbon investment would thus become part of capitalist development as usual in the neoliberal era or part of what Harvey (2005) describes as neoliberal project.

[177] In that sense, it is slightly different from related concepts like the post-fordist state in regulation theory that gives greater emphasis to the sites of production and the relative immobility of production processes for the influence and strategies of states in the globalization process (Cerny and Evans 1999).

perceive as the challenges of globalisation. The competition state, in that sense, 'is itself shaped by neoliberal state actors, often in the teeth of traditional left-right divisions and backlashes' (Cerny 2008: 34).

The transformation of the state also changes the power relations between state (and non-state) actors, enhancing the influence of some while limiting the role of others. The focus of the respective analyses often is on how central banks and finance ministries obtain a more important role through the increasing integration of economies in the world market.

Part of this internationalisation of the state is the growing importance of international, or rather transnational, coordination processes. Cerny (2008: 34) highlights the importance of transgovernmental networks of politicians and bureaucrats who interact and exchange on an increasingly regular basis: 'For example, financial regulators are likely to have far more in common with their interlocutors in other financial regulatory agencies [...] in terms of norms and understandings of how to deal with financial globalization, than they will with politicians and bureaucrats in other parts of their own state apparatuses'.

New, networked forms of governance require greater openness and enhanced interaction between state- and non-state actors; usually, the focus is on the domestic policy context where these forms of exchange are part of the creation of public-private partnerships, for instance. However, interaction and cooperation with non-state actors increasingly play an important role beyond the borders of the nation state as well, and are crucial in pursuing international competitiveness. The strategic importance of attracting foreign investment requires a greater openness for the concerns and interests of potential investors.

These forms of exchange and mutual learning processes play a crucial role in governing investment for climate protection as well. These exchanges often take place during thematic international conferences. A neat example for this is a panel during the 2010 Committee on Forestry meeting of the UN Food and Agricultural Organization (FAO), that aimed at *Communicating the potential of forestry to the finance sector*. In fact, the objective was to communicate interests and concerns in both directions, as the chair of the panel, Jerker Thunberg from FAO, highlighted: 'We foresters don't understand almost anything of the finance issue and vice versa'.[178]

Making forestry administrations and investors understand each other is thus a translation process in the proper sense of the word, as Emmanuel Ferreira, economic advisor to the government of Paraguay and one of the panellists, observed. When he first brought the two constituencies together, 'the forestry peo-

---

[178] All quotes own documentation: FAO Committee on Forestry, 5 October 2010, FAO Headquarters, Rome.

ple said we need long term loans with low interest rates, and the conversation was over at that point.'

As a reaction, Ferreira and his colleagues started to teach bank staff on forestry issues, with the result that some banks designed specialised financial products; vice versa, they tried to explain foresters how financial markets work, and how domestic regulation affects investment returns and cash flows. The crucial challenge for forestry departments that seek finance from capital markets, as Reinhold Glauner, Managing Director of WaKa Forest Investment Services explained during the same panel, 'is to create reasonable financial returns of about 10 per cent to attract the billions of dollars available from private and institutional investors'.

6.3.2   An emancipatory approach to the low-carbon transformation

The question that arises here is if and how the need to compete for foreign investment helps to organise the low-carbon transformation in a way that contributes to the wider social objectives that are often associated with it, like the democratisation of energy systems or changes to economic power relations, both within and between nation states. Formulating climate protection and low-carbon transformation strategies in a way that attracts the 'big pots of money' and allows for economic 'winners' rather suggests a focus on the concerns and interests of those who already are among these winners today.

Concerns in this regard have also been raised in the climate investment discourse, for instance with respect to using public climate funds to leverage private investment: 'Lower-middle income and least developed countries worry that their relatively weaker investment climates would deter private investment and leave them without access to funding. [...] [O]ther developing countries worry that local private sector actors will be cut out of business by large multinational financial concerns' (Sierra 2011: 2).

Such concerns do not remain unanswered. In the run up to COP 17 in Durban in 2011, an expert group on the governance of the Green Climate Fund (GCF) made proposals for aligning the concerns of developing countries with incentives for enhanced private sector engagement. According to these proposals, all operations of the Fund's private sector facility should follow country-driven mitigation and adaptation plans, promote the participation of local businesses, and prioritise private sector engagement in those countries that are the most affected by climate change and receive little FDI.

The GCF addresses the needs of different countries therefore. The fund's private sector facility 'provide[s] differentiated tools and approaches for coun-

231

tries that are still developing an enabling environment, for climate compatible investment, but also focus on business models which meet the needs of countries with stronger investment climates' (Sierra 2011: 2).

The government practices that form the emerging investment dispositif in climate governance address similar concerns. CDM reform proposals or the World Bank Climate Investment Funds pay attention to the different needs of countries in attracting foreign investment. The World Bank and other international organisations support these countries through capacity building programs and support in creating conducive investment frameworks. The discussions on a REDD mechanism for forest protection also considers the challenges for different countries and world regions to benefit from forest investment.

The climate investment dispositif, apparently, creates a web of regulation, incentives and support mechanisms that address environmental, social and development concerns related to clean investment. Through the lens of ecological modernisation, this can be read as learning processes and increasingly 'smart' regulation that gives greater concern to specific country needs, as opposed to the one-size-fits-all approach of the Kyoto Protocol.

Using the competition state lens, to the contrary, suggests that this framing obscures the pressures that the financial system exerts on the low-carbon transformation. The government of investment for climate protection thus is similar to what Cerny (2008) describes as reaction to the global economic and financial crisis: the creation of regulatory systems at the national and international levels that adjust domestic economies and economic policy making to global financial markets (Cerny 2008).

What is made invisible and neglected that way are forms of achieving the low-carbon transformation that do not follow the logic of the competition state and enforce its hegemony. To make sure: Formulating these alternatives is no easy task for individual states. 'As long as the concept of national [...] competitiveness remains in currency', Dickens suggests, 'then no single state is likely to opt out' (Dicken 2003: 141).

The pursuit of competitiveness is embedded within a powerful neoliberal rationality that also proved resistant to fundamental change in the recent economic and financial crisis. The possibility for states to become something other than a competitive entity is therefore 'likely to depend also on more general dehegemonisation of neoliberalism as a rationality of government' (Fougner 2006: 184).

Certainly, low-carbon strategies alone cannot remove the need to engage as competition state. Nevertheless, the understandings produced in these strategies are highly critical for these struggles, as they either challenge or naturalise the constraints imposed by global financial markets. Strategies for a low-carbon transformation can either help maintaining the image of a highly fragile but simultaneously powerful financial system that must be adapted to the climate challenge through a series of technical interventions and innovations; or contribute to a more radical problematisation of the current global financial system that would highlight how it is fundamentally ill-suited to address urgent global problems like climate change, not last by impeding cooperation and encouraging competition between states.

The problem with the climate investment discourse, in that sense, is that the focus on the financial and economic opportunities related to climate protection and the low-carbon transformation leads to a one-sided perspective on the investment challenge.

The Carbon Disclosure Project (CDP) is a good example for this. The project builds on an early initiative of the UNEP Finance Initiative to benchmark the CO_2 emissions of companies. It seeks to convince companies to voluntarily disclose their carbon intensity, and also to report on strategies to reduce this intensity in the future. The CDP was launched with great enthusiasm and the expectation that it would turn the carbon emissions of companies into a factor for investment decisions, and thus would allow using the power of investors for going low-carbon (Pattberg and Stripple 2008).

In itself, the CDP is highly successful, as both the number of investors who support the initiative and the number of companies that respond to the questionnaires has increased sharply. The problem, however, is that high carbon intensity thus far hardly lowers the value of a company, due to the lack of a (high) price on carbon (Newell and Paterson 2010). The prospects for change in this regard are highly uncertain as well.

To make the transparency provided by the CPD a factor in investment decision would require financial pressure on carbon emitters—it is difficult to imagine this pressure come from anywhere else than state intervention. Paradoxically, however, initiatives like the CDP are embedded in a discourse of self-regulation and non-hierarchical forms of governance that can replace what is perceived as old-fashioned top-down regulation.

In this discursive context, the space for bold environmental or climate change regulation remains limited, despite the increasing evidence that (ecological) regulation must not constrain economic prosperity and competitiveness. The climate investment discourse strengthens the image of powerless states that need to adapt their regulation to the interests of a powerful financial sector, rather than highlighting the many ways in which the financial sector depends on state activities vice versa. When national governments or the World Bank issue green bonds that guarantee maximum security for investors and returns above usual government bonds, it is not necessarily the interest to 'actively make "a green investment' (EIB 2009) that makes investors purchase this bonds, but simply good business sense (see chapter 5.1.3).

The bottom line of this could be that there is much more scope for states to actively shape the conditions of the low-carbon transformation than the current investment talk in climate governance suggests. Investor organisations emphasise that there is everything but a scarcity of investment capital, and also that there is growing concern for the security of investments. The global financial crisis has enhanced these problems. The provision of stable investment frameworks, let alone guarantees from public authorities, could thus be seen not only as 'tailor made risk mechanisms', but also as a way of creating more space for orchestrating the low-carbon transformation.

Alternative positive stories of climate protection and the low-carbon transformation that highligh the many benefits beyond *economic* or *financial* gains could contribute to creating this space for government choices and, most importantly, for more participation and democratic decision making in orchestrating the low-carbon transformation. There is no lack of evidence for these benefits: Traffic concepts that reduce not only carbon emissions but also noise and air pollution and improve urban life are one example here, decentralised, locally organised and controlled (renewable) energy systems are another one.

To be clear: These transformation processes cost money as well, but their realisation does not necessarily go along with opportunities for profit-seeking investment. Framing the low-carbon transformation as the pursuit of more healthy and satisfied (and potentially less unequal) societies would allow exploring alternatives to planning and financing the low-carbon transformation along the requirement of financial markets.

This does not require to reinvent the wheel, to be sure. One among many options ia a combination of Feed-In-Tariffs for clean energy supply and carbon taxes (possibly very low in the beginning but rising steadily). These would provide regulatory certainty and pressure on investment decisions on the one and, and the financial space for putting the interests of the population at the centre of decision making on the other hand.[179] Coupled with international financial support, such a strategy would also give developing countries greater flexibility and independence from capital markets in pursuing low-carbon development.

---

[179] To a certain degree, these instruments were introduced in Germany, but without making them part of a more emancipatory approach to the low-carbon transformation.

7 Conclusion

This book asked whether and how the increasing awareness for the economic significance of climate change contributes to and reflects a shift in the rationality of governing the climate. To answer this question, the book first sought to clarify what is different between the economic perspective on climate change brought forward by the Stern Review, and what has been described as market-driven climate regime since the adoption of the Kyoto Protocol. The main contribution of the Stern Review, it was argued, is to higlight the magnitude of the economic risks and opportunities related to climate change, and the need for and benefits from strategies that support a fundamental transformation of economies.

The climate investment discourse inherits much of this understanding. However, it is neither a necessary consequence of the Stern perspective nor its concrete expression. Rather, it is a particular strategy for addressing the economic challenges related to climate change that has developed in recenty years within several contexts. It coheres around the objective to drive (private) investment to the most urgent and effective climate protection activities.

How this effects on climate governance is best expressed in the changing understanding of the cost-effectiveness principle. Whereas the Kyoto approach to climate protection pursued cost-effectiveness through markets and flexible solutions, the climate investment discourse rationalises this objective and makes it accessible for government strategies. The analysis of Public Finance Mechanisms, the reform of the CDM, and the creation of a REDD mechanism confirmed that states obtain greater responsibility in orchestrating a low-carbon transformation: It is in that sense that we can speak of changes to the rationality of governing climate change, from reducing emissions to building low-carbon economies.

The last part of the book therefore reflected the new role of the state in climate governance and how it affects the understanding of the market-state relation in climate governance. It argued that the emergence of the climate investment dispositif reflects an uneasy compromise between the need to engage financial markets for the low-carbon transformation, and the limited capability or willingness to change the organisation of financial markets fundamentally. In the context of a financialised world economy, the state returns as competition state to climate governance therefore.

235

The climate investment discourse therefore fits into a renewed debate on the appropriate role of the state in the economy, which was triggered by the global financial crisis in particular. In both cases, however, the mounting evidence for the negative effects of powerful financial markets has not led to serious challenges of this powerful role, but to the creation of governance, support and security mechanisms that improve the stability of the system, and contribute to the reliability and attractiveness of domestic investment environments.

After this very quick summary of the main argument of the book, the primary objective in this conclusion is first, to reflect the usefulness of a discourse theoretical perspective for addressing the questions raised in this book and the difficulties that the research process encountered (7.1); and second, to use the findings from the analysis of the climate investment discourse for a look forward, at the prospects of international climate politics (7.2).

7.1 A look back: What has been achieved, and what remains to do

Two aspects were decisive for choosing a discourse theoretical approach for this book: First, to describe the link between an economic perspective on climate change and the investment discourse not as a causal relation, but as two expressions of the same discursive shift; and second, to highlight how discourses and practices constitute the objects they address.

The choice of theoretical concepts and research methods followed these objectives and largely enabled addressing the questions that were raised in the beginning. However, the research process also revealed a number of challenges. I will focus here on two issues that have been addressed throughout the book and remain critical for the contribution that poststructuralist methodologies can make to social science research.

On the one hand, these are methodological challenges. The introduction to this book identified in particular two questions that are raised with respect to the study of governmentality: How to account for the emergence and change of discursive or governmental constellations; and how to not lose sight of the diversity – and sometimes inconsistency – of governmental practices within the necessarily generalising perspective on governmental rationalities. The following reflects how combining the dispositif and regime of practices concepts allowed addressing these questions, but also highlights the challenges that remain, both with respect to the use of this concepts and the analysis of the climate investment discourse.

On the other hand is the question what contribution a poststructuralist research perspective can make for the evaluation and critique of policies and governmental practices, and whether it can contribute to their development that way. The question, in other words, is how the necessary distance of discourse theoretical research from the 'messy actuality' of governmental practices enables and constrains critical appraisals and recommendations.

7.1.1 Theoretical and methodological challenges

The book built on poststructuralist ontology to identify, explain and analyse changes to the approach taken to climate protection, and combined the concepts problematisation, dispositif, practices and logics in a research strategy that integrates genealogy and archaeology. The dispositif perspective in particular allowed addressing several of the challenges related to the study of governmentality.

Explaining emergence and change of governmental formations

For the *archaeological dimension* of the research, this book built on the logics concepts that Howarth and Glynos (2007) introduce for the study of regimes of practices, to capture what makes a certain set of practices 'tick' (XY). However, while they give great emphasis to distinguishing social from political logics, they offer little advice on how to actually identify and define these logics in the research process.

It was therefore argued that the logics concept describes an analytical middle ground between a nominalist perspective that focuses on the formation of concepts, and the much broader perspective of governmental rationalities: In that sense, logics describe common patterns of governmental thought or knowledge that specify government rationalities but are not necessarily reflected (yet) in common terminology.

This understanding allowed identifying and describing an investment dispositif in climate governance that links issue areas such as climate finance, low-carbon strategies and renewable energy policies, and instruments or governmental technologies like PFMs, carbon markets and the REDD mechanisms—though this link is usually not made explicit in the practices of those who address these issues.

The dispositif concept also allowed to address one fundamental challenge in the study of governmentality. Governmentality studies are sometimes accused of focusing on the systematic nature and internal coherence of discursive or governmental constellations, but not to account for their emergence and change. In that sense, they seem to have a structural bias and lack an agency perspective.

Such critique, however, 'ignore[s] governmentality's genealogical foundations and thus its emphasis on the contingent and invented (and thus always mutable) nature of governmental thought and technique' (Rose, O'Malley et al. 2006: 99). We saw that Foucault highlights – through the notion of problematisation – both thinking as a driver of historical change and the (structural) conditions for governmental thought and practice. Governmental practice takes place within these conditions but is not their necessary consequence; rather, it is an attempt by governmental agencies to make sense of and cope with particular social conditions.

The dispositif concept is well suited to capture this interplay of structural conditions and agency. As analytical tool or *grid of interpretation*, the dispositif helps to account for the emergence of practices and institutions within certain rationality or discursive constellation, but also highlights how this rationality formed from these practices vice versa (Howarth 2000). It highlights how the strategies of different actors cohere around a common strategic objective, without the need for a hegemonic force or a congruence of interests between all of these actors. Rather than to ask who put through his interests, dispositif analysis examines the contexts and practices that contribute to the emergence of a problematisation, and how it is accessible for a variety of strategies and interests.

Accordingly, genealogy was interpreted here in a way that requires including a broad set of empirical material, to explain from different perspectives how a variety of processes contribute to the emergence of investment as an object of government.

Genealogy, as an analytical perspective, also requires certain methodological openness therefore. It is in this sense that Foucault suggests that a genealogical style of explanation replaces strong causality with a pluralisation of causes, and leads to an 'increasing polymorphism' of elements and the relations between them (Foucault 1991: 77). Along the same lines, Dean (2010: 39) argues that analysing the emergence of governmental practices means to 'examine all that which is necessary to a particular regime of practices of government, the conditions of government in the broadest sense of that word. In principle, this includes an unlimited and heterogeneous range of things'.

Clearly, as Dean also highlights, the analysis does not simply list these things, but reflects how they are rationalised in governmental thought and practice. Nevertheless, genealogy refuses methodological formalisation to a certain degree, a fact that is sometimes confused with a lack of explanatory power. A genealogical style of analysis, Foucault proposes, 'is at once too much and too little. There are too many diverse kinds of relations, too many lines of analysis, yet at the same time there is too little necessary unity' (Foucault 1991: 78).

Getting closer to the messy actuality of government practice

One added value of the dispositif perspective was thus clearly to highlight how various domains of government are linked through a common strategic objective, without claiming that the practices in these domains are determined by the strategy. Rather, the dispositif perspective understands this as coherence around increasingly common but not necessarily reflected objectives: CDM reform proposals and the creation of PFMs, in that sense, both aim at governing investment for climate protection, without making explicit this common ground.

The question that arises, however, is how to handle this diversity of practices in the further analysis, in particular with respect to the interpretation and critique of these practices. The second methodological and analytical challenge confronted in this book was thus to handle the tension between the objective to identify a common rationality or strategy that informs and drives the practices directed at investment for climate protection, and the need to not lose sight of the differences and specifics of these practices.

In a very practical sense, this challenge became visible in the first step of the research, that is, in identifying, delimiting and describing the climate investment discourse as an object of research. To a certain degree, the difficulties related to this task were an issue of timeliness: The turn to finance and investment only occurred while this book was written, and to a certain degree, its importance caught me by surprise. Initially, the book set out to understand and explain how the discussion around the publication of the Stern Review established a certain economic rationality in climate governance, and what the effects of this rationality are.

Soon it became clear, however, that the many different ways in which a variety of actors started to address investment for climate protection was related to this question in important ways, and that the government of investment provides the most palpable object for studying change to the economic rationality in climate change governance.

But even after the finance and investment turn became the main empirical object of this book, the investment discourse kept growing too fast and in far too many directions to look at all its new branches. One consequence of this, as should have became clear throughout the previous chapters, is that despite the growing consensus on the need to address investment for climate protection, there are still many different interpretation of this challenge and strategies for addressing it.

This highlights a challenge for the analysis that is not exclusive to this book: How to identify the common logics or rationality that drive the emergence and development of a domain of government, while paying attention to the differences between the practices within this domain.

Once again, the dispositif perspective offers a promising way forward here, as it does not claim that the practices directed at investment for climate protection are determined solely by the objective to raise finance or investment; rather, their concrete form depends on the actor constellations, interests, institutional legacies and rationalities that already existed in the contexts in which they develop. While, for instance, proposals to reform the CDM and the creation of a REDD mechanism are clearly affected by the climate investment discourse, the concrete developments and changes we see here are also largely influenced by the institutional legacy that was put in place with the adoption of the Kyoto Protocol.

The dispositif perspective thus allows identifying the common ground in a variety of processes and practices, while leaving enough space to enquire deeper into the specifics of some of these processes. However, the need to find a middle way between analytical depth and a broad and inclusive perspective on the constitution of investment for climate protection as an object of government means that several important issues have remained unconsidered in this book. I will focus on two possible routes for further enquiry here.

Though chapter 5 aimed at describing thoroughly the practices that aim at raising and directing investment for climate protection and the low-carbon transformation, many questions related to the government of investment remain unconsidered here: How do the institutions that govern investment for climate protection interact with investors, and how are priorities formed this way? How do perceptions and understandings of the climate challenge emerge from and change in these interactions on both sides? And through which tools and forms of analysis do investors incorporate the incentives provided by states into their decision-making processes?

A second and related issue that the book could only touch upon briefly is the constitution of investor groups as a new actor in climate governance, and the processes of identity formation that go along with this. Descheneau and Paterson describe how carbon traders combine narratives of environmental or climate protection with existing financial practices, to constitute themselves as actors in the carbon markets (Descheneau and Paterson 2011). In a similar way, more analysis is required to describe the way the climate challenge is taken up by the so-called investment community; how the need to address a highly political issue like climate change effects on the self-understanding of investors; and whether this brings about forms of investor behaviour that go beyond *finance-as-usual*.

Addressing these and related question would be a way of enquiring much further into the 'messy actualities' (Rutherford 2007: 505) of governing investment for climate protection than was possible in this book, and would contribute to understanding how investment for climate protection is constituted as an object of government.

7.1.2 Discourse theory analysis and the question of critique

Another important issue related to the choice of the analytical approach in this book is if and how it is possible to make, on basis of a discourse theoretical analysis, a critical appraisal of policy formations and governmental practices, and also recommendations for change. What is at stake here is best reflected in a question that I was repeatedly asked, albeit in different versions, when presenting my research to an audience concerned with climate change governance (rather than with the challenges of discourse theoretical research designs): 'So if you had the opportunity to meet the World Bank staff that is responsible for their climate investment activities, what would you recommend them to do?'

The first reaction to this question was usually to remind that the objective of discourse theoretical research is not to come up with a set of applicable policy recommendations, and that poststructuralist ontology is not only inadequate for this objective, but essentially aims at problematising this form of advice. Certainly, this answer is justified from a scientific point of view; nevertheless, the question remains what insights a book concerned with urgent political challenges like climate change from a discourse theoretical perspective can – and should – provide for addressing these challenges.

No doubt, this question must be considered with all the caution imposed by poststructuralist ontology. Nevertheless, as Howarth and Glynos (2007) remind us, any (social science) research project entails a certain normative claim, if only for problematising a particular issue or object; the choice of the analytical object already suggests that it is worth questioning current understandings, and that alternative understandings could contribute to a better way of governing this issue or object, whatever this means in concrete terms.

It is thus not necessary to make a bold judgement of the government of investment or detailed proposals for its improvement: if the task is to identify ways of enhancing the efficiency of climate finance mechanisms or the governance of carbon markets, the perspective chosen in this book is indeed not the most promising one. However, through a critical appraisal of what is made visible in the climate investment discourse, and by also highlighting important aspects that are made invisible within this framing but deserve more attention in the political process, it can make a very practical contribution to debates on financing climate protection and the low-carbon transformation.

It is possible to highlight, in that sense, several aspects of the turn to finance and investment that allow a better understanding of the development of climate governance in recenty years. The investment discourse reflects, first of all, the failure of a strategy that built narrowly on (carbon) market forces for achieving climate protection, and marginalised to the same degree the role of states and regulation.

This is reflected in many ways in this book: Public Finance Mechanisms highlight the need for more state intervention and support in directing investment to the desired ends; the reform of carbon market mechanisms aims, explicitly or implicitly, at higher regulation of carbon finance flows; and discussions on a REDD mechanism highlight concerns with purely investor driven markets for forest protection.

Second, the climate investment discourse raises the awareness for the diversity of challenges related to climate protection and a low-carbon transformation. It thus challenges the Kyoto approach to climate protection that concealed the diversity of challenged in the abstracted unit of emission reductions, and let market forces chose from a broad set of emission reduction opportunities. Climate protection after the finance turn, therefore , ascribes a more active role to the state in identifying and realising the changes necessary for a low-carbon transformation.

Third, the climate investment discourse brings to attention the role that investors and financial markets play in combating climate change. In principle, this awarenes allows for exploring new ways for meeting the climate challenge, in particular through framing climate protection as a positive story that brings to the light the (economic and social) benefits from going low-carbon. While the emphasis on economic opportunities is not without caveats neither (see below), it could in general help to lower the influence of those who slowed down progress in climate governance for many years, by cautioning against the negative economic consequences of climate protection.

However, this book also revealed how the climate investment discourse limits the scope for political intervention, and contributes to a narrow perspective on aligning ecological and economic objectives.

First of all, if there was any doubt left that the dominant response to the climate challenge would have at its heart the development and deployment of (new) technologies, the finance turn in climate governance has finally dispelled these doubts. It is technological change that can be achieved through finance, and, as Nicholas Stern and others emphasise, investment in new technologies and industries could spur a new cycle of growth. While no succesfull strategy for combating climate change can do without new technologies, the climate investment discourse manifests a pro-growth orientation and contributes to further marginalising all forms of climate protection that are not technology driven, such as the change of lifeststyles.

Second, the awareness for the magnitude of the climate challenge and the resulting need for new climate protection stategies has not led to a politicisation of climate governance, if we understand politicisation as the processs of foregrounding the causes of a problem and a broad discussion about possible changes and alternatives. All to the contrary: against the economic and finanical challenges related to climate change, it has become even more common to advocate expert interventions and technical solutions for raising the required levels of finance and creating conducive environment for low-carbon investment.

This is best reflected in how the notion of (investment) market failure is key for claiming a new responsibility of the state in addressing climate change, and for limiting its role at the same time. The understanding of climate change as biggest market failure in history does not lead to considering the dominance of markets as part of the problem, but to reducing the problem to a mere technical challenge, and to detaching the *failure* from the other – positive – characteristics of markets.

The risk terminology is the discursive ally of the market failure notion in that sense, as it helps to insulate certain market features from change. The role of a Green Investment Bank in the UK, in that sense, would be to reduce 'unacceptable market risk' (GIB 2010: 13): Whereas governments are expected to create risk-reward relations that meet the expectations of investors, the understanding of what are acceptable risks or returns, and the implications of these expectations, are not part of the discussion.

A third problematic understanding entailed in the climate investment discourse is that developed and developing countries increasingly face the same challenges with respect to achieving climate protection. The need to raise or attract investment for the low-carbon transformation means that countries with lower levels of FDI need to adapt their investment frameworks and institutional settings to those that have been more successful in that regard. This equalisation complicates critique of how the approach to investment for climate protection manifests capitalist logics and the need to act as competition state: Questioning strategies to attract clean investment to developing countries would mean to also question what is described as an important opportunity for economic development.

Altogether, there is the need for broadening debates on the right strategies to address investment for climate protection and the low-carbon transformation; what is required in particular is a broader perspective at the many ways in which financial market regulation and climate change governance are intrinsically linked. Strategies for the decarbonisation of economies cannot be at the centre of addressing – and let alone overcoming – the many challenges and constraints that powerful financial markets impose on the low-carbon transformation and economic policy making more in general—nevertheless, they can make an important contribution to highlighting these challenges.

7.2 A look forward: What future for international climate politics?

There is little reason for doubt that the government activities described in this book will continue and play a more important role in climate governance in the future. Efforts for enhancing renewable energy supply or the production and deployment of clean technologies will become priorities for states, as the economic opportunities related to these developments become more visible and, equally important, increasingly independent from climate regulation; rising energy prices and concerns for energy security are likely to further trigger the interest in low-carbon economic development.

The question arises, however, what role remains for the internationally coordinated efforts for climate protection. The introduction to this book, in that sense, raised the question why the widely shared insight that climate protection is rational from an economic point of view has not translated into political action, and whether this particular economic problematisation of climate change effects on the development of international climate politics at all.

The analysis of the climate investment discourse offers an interesting perspective to address these questions: The focus shift to the domestic dimension of climate protection matches with recent claims that question the feasibility (or desirability) of coordinated international efforts to combat climate change, and suggest replacing these with a multilayered governance architecture. The following builds on the analysis of the climate investment discourse to challenge these arguments to a certain degree, and to propose an alternative perspective on the role of international cooperation for climate protection.

7.2.1 What is wrong with a global top-down approach

When describing the emergence of low-carbon growth strategies, chapter 4 offered two interpretations of the role that these strategies could play for the development of international climate change politics: Either, they could be a way of realising internationally agreed targets and thus enhance their effectiveness; or increasingly replace these and reduce the international dimension of climate governance to voluntary cooperation and exchange on best practices.

Recent developments strongly support the second interpretation. The years since the adoption of the Kyoto Protocol provided ample evidence for the difficulties in achieving meaningful international agreement; this, apparently, contributes to a fading belief in the feasibility of halting global warming through international cooperation more general.

COP 15 in Copenhagen in 2009, in that sense, was the climax of a form of international climate diplomacy that aims at the global management of climate change. Despite the urgent need to agree on a political framework for the Post-Kyoto time, and the great expectations that were raised in the run-up to the Copenhagen summit, governments could not achieve meaningful steps forward. 'The gap between the pre-Copenhagen rhetoric of "what must be done to stop climate change" and the reality of the Copenhagen Accord outcome was spectacular. [...] No agreement of much consequence was reached, and the very efficacy of multilateral climate diplomacy through large set-piece conferences was called into question' (Hulme 2010).

The developments since Copenhagen hardly allow for a more optimistic reading: Accompanied by lower expectations, COP 16 in Cancún in 2010 achieved no breakthrough for international cooperation, and COP17 in Durban one year later also ended in disappointment, at least when measured against the urgency given to quick and substantial progress: While states agreed to enter into another commitment period of the Kyoto Protocol and to create a more encompassing, legally binding instrument from 2020, relevant details like the envisaged level of emission reductions and the legal status of the planned new agreement remained open once more.[180]

Not much more progress could be made on the issues that seemed much closer to solution prior to the summit, like a forest protection mechanism or the Green Climate Fund: Whereas, in case of the latter, Parties decided to initiate a

---

[180] German chancellor Angela Merkel upset climate diplomats and activists in the early days of COP 17 by saying that agreement on the continuation of the Kyoto Protocol is not likely. As if to support this pessimism, Canada left the Kyoto Protocol days after the Durban summit and thus put in question the agreement achieved here, as China and other emerging economies had agreed to binding emission reductions from 2020 only in combination with a continuation of the Kyoto Protocol.

work programme to make the fund operational, neither the financial sources to bring up the envisaged US 100 billion annually from 2020 nor the level of funding for the years until then could be agreed upon. The only concrete decision was to not use a levy on aviation and maritime travel for climate funding.

Against this background, voices are getting louder that a global top-down approach to climate protection is neither feasible nor desirable and should be replaced by a bottom-up approach, to allow for a greater diversity of measures and initiatives. The 'undeniable and highly public collapse of the Kyoto route has opened a space within which open-minded reappraisal of the basic policy architecture is possible for the first time in over a decade' (Rayner 2010: 617).

Prins and Rayner thus propose 'a radically different approach from the top-down command and regulatory regime of output targets that is Kyoto' (Prins and Rayner 2007b: v). Their objections against the past direction of international climate politics cohere around the Kyoto Protocol as a universal intergovernmental treaty, and especially the objective of reducing emissions through the creation of a global market and a single carbon price signal.

While they do not reject the idea of a global carbon price to spur technology innovation in general, they nevertheless raise two major objections: First, that 'to work, such schemes must be built – like all genuine markets – from the bottom up' (ibid 974); and second, that a cap-and-trade schemes cannot give incentives quick and strong enough to push investment for technological innovation in the time needed. They thus propose a dramatic increase in public investment for technology research and development instead.

Their more general objection against a global top-down approach to climate protection is that the majority of changes for climate protection and the low-carbon transformation are required at the national or even subnational levels anyway. 'There may well be some aspects of policy that require international agreements among many countries, but the basic assumption here is that these instances are far fewer than conventional climate policy supposes and that any such an agreement is neither the starting point nor a precondition for effective climate policy' (Rayner 2010: 617).

A related critique to a global approach to climate protection that is backed by the findings in this book is that framing climate change as a global challenge that requires global solutions is an important dimension, if not a driver, in the preference for expert-driven technical solutions, and allows for the abstraction from the concrete spatio-temportal contexts of climate change drivers and solutions. Shifting climate protection strategies from the global to lower spatial levels wherever possible could contribute to creating space for more democratic decision making processes in climate change governance therefore (Methmann 2012).

The claims for a multilayered approach to climate change are backed by an increasing number of climate protection initiatives beyond the global negoatiations. Among the most prominent examples are the voluntary Asia-Pacific Partnership on Clean Development and Climate (APP) that focuses on technology cooperation rather than emission reduction agreements; local level cooperation like the Cities for Climate Protection Campaign; and private sector initiatives like the CDP and cooperation between businesses and NGOs.

The proliferation of partnerships, local initiatives and public-private cooperations has led to a fragmentation of (global) climate governance or a 'regime complex' that comprises, beyond the UNFCCC, the work of UN agencies, Multilateral Development Banks and other bilateral and unilateral initiatives (Keohane and Victor 2010). This is not seen as a problem necessarily: 'While some nations hope to maintain a universal approach towards climate governance, others seemingly work towards new forms of a more fragmented and flexible order that places more emphasis on hybrid and private mitigation policies' (Pattberg and Stripple 2008: 368). It is in particular the potential of multi-stakeholder partnerships that raises positive expectations (Biermann, Chan et al. 2007).

The fragmentation of climate change governance is neither seen as an obstacle to progress in the international negotiations necessarily, as a blog post by Andrew Steer, World Bank Special Envoy for Climate Change, during the Durban climate talks in December 2011 suggests: 'Perhaps the greatest hope comes from the thousands at COP 17 who won't be negotiating UNFCCC text. These are the private sector, civil society, researchers and international organizations. Their job is to move forward the world of action, sharing experiences and analysis, launching new programs, and doing deals' (Steer 2011).

However, Steer continues, this world of action should not replace the world of negotiations; to the contrary, it could become a driving force for agreement: 'The pace of innovation on both adaptation and mitigation is remarkable – and now needs help from internationally agreed rules in order to scale up. The more the negotiators look over the shoulders at this group, and see that success is possible and affordable, the better the prospects will be' (idid.).

The hope that Steer addresses here is that once states realise the many benefits related to addressing climate change, agreement should become more feasible. This, once more, stresses the positive narrative of climate protection: The economic opportunities related to climate change should not only help to realise individual climate protection projects or nation-wide efforts then, but also describe a way forward for international agreement.

It is in this sense also that Prins and Rayner (2007b: 975) support the idea of a *global federalism of climate policy*: 'Rather than the top-down universalism embodied in Kyoto, countries would choose policies that suit their particular circumstances. [...] Although a bottom-up approach may seem painfully slow and sprawling, it may be the only way to build credible institutions that markets endorse'

7.2.2   But what still might be right about international cooperation

Before ditching the Kyoto approach to climate protection, however, and with it any aspiration for meaningful international agreement on climate change, it is worth to qualify and reflect the objections made against a global treaty on climate protection, and to clear up some misunderstandings in this regard.[181] The findings in this book offer a good background for this.

Like others, Prins and Rayner criticise that the Kyoto approach was modelled on the experiences with instruments like the Montreal Protocol and a trading scheme for sulphur emissions in the US, but failed to recognise that climate change is a wicked problem that cannot be solved by focusing on a singly object like carbon emissions. Hulme (2010) therefore laments that climate change 'has been represented as a conventional environmental "problem" that is capable of being "solved." It is neither of these. Yet this framing has locked the world into the rigid agenda that brought us to the dead end of Kyoto [...]'.

The objections against a global approach to climate change cohere around three related aspects, then: The (narrow) focus on emission reductions; the pursuit of a legally binding agreement or treaty; and climate protection through a single carbon price achieved through a global carbon market. So while the Kyoto Protocol 'is a symbolically important expression of governments' concern about climate change, [...] as an instrument for achieving emissions reductions, it has failed' (Prins and Rayner 2007a: 973).

No doubt, Prins and Rayner raise important issues here, and history seems to prove them right: The Kyoto architecture has not led to substantial emission reductions and neither spurred much innovation in renewable energies or clean technologies. The failure of the CDM and the other flexibility mechanisms in this regard has been discussed in this book.

But we need to have a closer look at the proposed alternatives to the Kyoto architecture to understand what has been wrong about a global top-down ap-

---

[181] I will focus mainly on contributions of Gwyn Prins and Steve Rayner here who have claimed that the Kyoto approach puts climate protection into the wrong trousers (Prins and Rayner 2007a; Prins and Rayner 2007b; Rayner 2010); and the Hartwell Paper that brought together a group of prominent British scholars to consider a 'new direction for climate policy after the crash of 2009' (Prins, Galiana et al. 2010).

proach to climate protection, and why international cooperation might still be right and necessary.

These proposals focus, on the one hand, on more immediate interventions to encourage technology innovation: public investment in research and development, in that sense, 'should be placed on a wartime footing' (ibid. 974). Hulme (2010) also suggests that '[s]ignificant public investment in direct decarbonisation of the global energy system is the most ambitious goal but is, in our view, more likely to achieve success than the existing illusory thinking about targets, timetables and trading'.

The other, more general suggestion is to replace a single, unifying approach to climate protection with a more diverse process, 'designed to generate a range of possible solutions, which can be compared and assessed, mixed and matched, changed and refined as we pursue the goal of climate security' (Prins and Rayner 2007b: 5). Hulme questions the viability of climate policy 'that has emissions reduction as the all-encompassing and driving goal', and advocates instead to accept 'that taming climate change will only be achieved successfully as a benefit contingent upon other goals that are politically attractive and relentlessly pragmatic' (Hulme 2010).

Hulme and others thus advocate what has been described in this book as a re-framing of climate protection, from reducing emissions to building low-carbon economies. We can see an important parallel between these proposals and the climate investment discourse therefore, in focusing on the – economic – benefits and the (political) feasibility of climate protection strategies. Massive public investment in low-carbon technology research is expected to meet less resistance than pricing carbon (globally) through either taxes or caps; at the same time, this investment can contribute to developing strong domestic companies and industries, and thus fits well into the strategic objectives of the competition state.

Prins and Rayner (2007a), in that sense, generally acknowledge the desirability of a carbon tax, but question its feasibility due to the difficulties with similar attempts in the past, in the US and elsewhere. Three years later, Rayner (2010) suggests a very modest carbon tax, but highlights that if 'the initial purpose of the tax is to raise revenue for RDD&D, it can be set at a level much lower than anything that would have a sustained influence on consumer behavior' (Rayner 2010).

The authors of the Hartwell Paper, a group of mainly British academics concerned with climate change, use more prosaic terms but actually mean the same when advocating that a strategy informed by political pragmatism is 'likely to be more effective than the approach of framing around human sinfulness - which has failed' (Prins, Galiana et al. 2010: 5)

What the climate investment discourse and the proposals to ditch Kyoto have in common, then, is that they highlight the objective to enhance efforts for climate protection and a low-carbon transformation in a way that neither threatens the economic interests of states or non-state actors, nor constrains the freedom of consumers. Compared to the Kyoto situation, we can see this as an important step forward, at least in one sense: Whereas the objective to protect economic and consumer interests was enshrined in the narrative of market-efficiency at that time, it is now put to the table in a more straightforward way.

This reminds us of the two different economic problematisations of climate change that chapter 6 of this book identified: Both the Kyoto approach to climate protection and the climate investment discourse aim at organising climate protection in a way that is beneficial to economic growth and competitiveness; but whereas the former builds on market flexibility to reduce the pressure on important economic sectors, the latter brings to mind the need for a transformational change to the economy, and to pursue these changes through more government intervention instead of leaving it to carbon market forces to choose from a broad set of emission reduction opportunities.

It is in this light that we have to weigh the arguments for climate protection strategies that are 'politically attractive and relentlessly pragmatic' and caution against putting pressure on consumer behavior. To make sure: There is nothing wrong with considering new ways of realising (and financing) climate protection and the low-carbon transformation, and the development and deployment of less energy intensive technologies in particular. Likewise, organising climate protection around the interests and concerns of state and non-state actors is more likely to gain support. What is critical, however, is the way that many of these proposals refrain from any form of regulation that might hurt somebody's economic interests.

It is correct that the lobby power and political pressure of powerful business actors has impeded the realisation of climate protection objectives in the past. Nevertheless, it is necessary to emphasise, as Nicholas Stern does (albeit with a lot of caution), that a low-carbon transformation will produce losers as well, as some sectors and industries will have to shrink considerably and ultimately disappear. Stern also highlights that successfully combating climate change will entail initial costs, from the changes that are necessary in developed countries and from supporting developing countries financially (Stern 2009a).

And the analysis of the investment discourse also showed that there are already a growing number of activities related to climate protection that states and non-states actors already pursue out of self-interest. Additionally, there are (potential) economic opportunities related to emission reductions that businesses cannot realise due to difficulties like the principal agent problem related to enhancing energy efficiency (see chapter 4.1). Domestic frameworks and incentive schemes can contribute to making these options available and to enhance, from a macro-perspective, the economic benefits of climate protection.

However, the analysis also showed that important dimensions of the low-carbon transformation are not made visible in an investment perspective: These are, on the one hand, all those activities that contribute to the climate crisis and still offer attractive incomes and returns on investment; and on the other hand the wider social and political objectives related to the transformation of economies that do not offer financial or investment opportunities immediately, but can contribute to the wellbeing of societies in more encompassing ways.

It is against this ambivalent diagnosis of the economic opportunities related to climate change that international cooperation would have to be reconsidered. We saw that the critics of the Kyoto approach are right in crucial regards. The focus on legally binding emission reductions was not very successful in the past and neither seems to be very promising for the future: these agreements have little substance and reproduce the burden sharing perspective of climate protection as a purely negative activity. Given this combination of non-effectiveness and negative framing, there are good reasons to abandon the Kyoto targets and timetables approach, and with it a strategy that bundles the many challenges related to climate change in a single object (carbon) and addresses it through a single strategy (a carbon market).

This, however, must not signify to give up the pursuit of international agreement on climate change altogether. What would be required instead of targets and timetables are forms of international cooperation that have more immediate effect on climate protection, both in forms of constraints and incentives. The authors of the Hartwell Paper make an important point here in claiming that '[w]ithout a fundamental re-framing of the issue, new mandates will not be granted for any fresh courses of action, even good ones'; they suggest that energy access, sustainability and resilience are the goals that would allow such reframing and give new momentum to climate politics.

However, they do not tell us where the incentives for achieving these goals are expected to come from. Whereas it is clear that the concrete changes for climate protection and the low-carbon transformation are required below the global level, a global framework could create space for these activities and also provide incentives. Otherwise, what is presented as a much more feasible approach to climate protection that harnesses 'enlightened self-interest' could turn into enhanced competition for low-carbon investment where financial gains can be made from emission reductions, and non-activity everywhere else.

Coming up with proposals for such a framework is not an easy task, as many possible ways forward have been disclosed due to the obsession with protecting domestic economies or particular constituencies against the negative impact of climate regulation. However, some of the old proposals deserve to be considered again with respect to how they combine constraints and incentives in pursuing climate protection.

To give one example: A carbon tax agreed on in Kyoto in 1997, even at a very low rate and only between the countries that were willing to do so, would have had more effect and given clearer signals for the low-carbon transformation than all targets, timetables and carbon trading systems did in the 15 years since Kyoto. What is more, the earnings from this tax would have allowed to realise climate related policies and to achieve the manifold (economic and social) benefits that are possibly related to this.

The objective here is by no means to describe carbon taxes as the solution to all problems related to climate change, and neither is a carbon tax a particularly radically approach to climate protection: it remains within the paradigm of market oriented climate governance. But international agreement on a carbon tax could nevertheless be a powerful symbol for the willingness to actively manage the low-carbon transformation process, and to explore new space for state intervention and cooperation.

The climate investment discourse with its sense of opportunity could contribute to open this space. The growing understanding that creating low-carbon economies will be an important factor for economic growth and prosperity in the future could facilitate agreement on common standards or coordinated taxes. Discussions and strategies would have to go beyond realising financial and economic opportunities quickly, however.

Instead of shifting climate governance from global to lower spatial levels only, it is suggested here to turn climate protection and the low-carbon transformation from a technical into a political challenge also. The authors of the Hartwell Paper suggest dedicating the earnings from a moderate carbon tax to an innovation fund that finances technology innovation exclusively, as this 'removes the issue from the political arena [...] and by doing so, may help to restore public trust at a time when the stock of politicians is not high in many of the democracies' (Prins, Galiana et al. 2010: 34).

However, this 'relentlessly pragmatic' approach is part of the problem, not part of the solution. Instead of keeping the decision on the use of carbon tax revenues out of the political arena, these debates should help to politicising the climate issue, domestically as well as internationally. Such discussions would highlight the benefits from using tax revenues and create space for realising some of the positive stories of climate protection than go beyond immediate financial gains. In that sense, international cooperation must no be opposed to climate related activities and decision-making processes on lower spatial levels neither. Global cooperation could enhance instead of delimit the freedom for policy choices and, much more substantially, enable a more emancipatory approach to the low-carbon future that is also the precondition for more democratic decision making in climate governance.

Bibliography

Ackermann, Frank (2004). Priceless benefits, costly mistakes: what's wrong with cost–benefit analysis? . In: Real-World Economic Review 25: 2-7.

Ackermann, Frank (2007). Debating Climate Economics: The Stern Review vs. Its Critics. Report to Friends of the Earth-UK. Medford: Tufts University.

Ackermann, Frank (2009). Financing the Climate Mitigation and Adaptation Measures in Developing Countries. New York, Geneva: UNCTAD.

AGF (2010). Report of the Secretary General's High-level Advisory Group on Climate Change Financing New York: United Nations.

AGF (2010). Report of the Secretary General's High-Level Advisory Group on Climate Change Financing. New York: United Nations.

Andrew, Jane, Mary Kaidonis, et al. (2010). Carbon tax: Challenging neoliberal solutions to climate change. In: Critical perspectives on accounting 21(7): 611-618.

Arrow, Kenneth (2007). Global Climate Change: A challenge to policy. In: Economists Voice 4(3): 1-5.

Atteridge, Aaron, Clarisse Kehler Siebert, et al. (2009). Bilateral Finance Institutions and Climate Change: A Mapping of Climate Portfolios. Stockholm Environmental Institute.

Bäckstrand, Karin and Eva Lövbrand (2006). Planting trees to mitigate climate change: Contested discourses of ecological modernization, green governmentality and civic environmentalism. In: Global Environmental Politics 6(1): 50-75.

Bailey, Ian and Sam Maresh (2009). Scales and networks of neoliberal climate governance: the regulatory and territorial logics of European Union emissions trading. In: Transactions of the Institute of British Geographers 34(4): 445-461.

Bales, Carter and Rick Duke (2009). Promoting economic recovery through climate legislation: Mc Kinsey. http://whatmatters.mckinseydigital.com /climate_change/promoting-economic-recovery-through-climate-legislation. Retrieved: 13.11.2011.

Barker, Terry (2008). The economics of avoiding dangerous climate change. An editorial essay on the Stern Review. In: Climatic Change 89: 173-194.

Beinhocker, E. (2006). The origin of wealth: evolution, complexity and the radical remaking of economics. London: Random House Business Books.

BERR, Defra, et al. (2007). Report of the Commission on Environmental Markets and Economic Performance. London: Department for Environment, Food and Rural Affairs.

Biermann, Frank, S. Chan, et al. (2007). Multi-stakeholder partnerships for sustainable development: Does the promise hold? , in: Pieter Glasbergen, Frank Biermann and Arthur Mol: Partnerships, governance and sustainable development. Reflections on theory and practice. Cheltenham: Edward Elgar, 239–260.

Blok, Anders (2010). Topologies of climate change: Actor-network theory, relational-scalar analytics, and carbon-market overflows. In: Environment and Planning D: Society and Space 28(5): 896-912.

Blyth, William, Ming Yang, et al. (2007). Climate Policy Uncertainty and Investment Risk. Paris: IEA/OECD.

BMU (2009). The International Climate Initiative of the Federal Republic of Germany. Berlin: Federal Ministry of the Environment Nature Conservation and Nuclear Safety.

Bond, Ivan, Maryanne Grieg-Gran, et al. (2009). Incentives to sustain forest ecosystem services: A review and lessons for REDD. London: IIED.

Boucher, Doug (2008). Filling the REDD Basket: Complementary Financing Approaches. Washington D.C.: Union of Concerned Scientists.

Boyd, Emily, Nathan Hultman, et al. (2007). The Clean Development Mechanism: An assessment of current practice and future approaches for policy. Tyndall Centre Working Paper 114. Oxford: Tyndall Centre.

Brand, Ulrich (2009b). Post-Neoliberalismus und der Staat. Zur aktuellen Debatte. In: Widerspruch 57: 93-101.

Brand, Ulrich and Nicola Sekler (2009). Postneoliberalism: catch-all word or valuable analytical and political concept? – Aims of a beginning debate. In: Development Dialogue 51: 5-14.

Bretton Woods Project (2008). 'No regrets': The World Bank and climate change Bank proposes vast expansion of its role, ignoring criticism of past record. http://www.brettonwoodsproject.org/art-562468. Retrieved: 20.02.2010.

Brinkman, Marcel (2009). Incentivizing private investment in climate change mitigation, in: Richard Stewart, Benedict Kingsbury and Bryce Rudyk: Climate Finance. Regulatory and Funding Strategies for Climate Change and Global Development. New York/London: New York University Press, 135-142.

Brunnengräber, Achim (2007). The Political Economy of the Kyoto Protocol, in: Leo Panitch and Colin Leys: Coming to Terms with Nature, Socialist Register 2007. London: Merlin Press, 213-230.

Bulkeley, Harriet, Matt Watson, et al. (2007). Modes of governing municipal waste. In: Environment and Planning A 39: 2733-2753.

Bumpus, Adam and Diana Liverman (2007). Accumulation by decarbonisation and the governance of carbon offsets. In: Economic Geography 84(2): 127-155.

Butler, Judith, Ernesto Laclau, et al. (2000). Contingency, Hegemony, Universality. London, New York: Verso.

Caldwell, Raymond (2007b). Agency and Change: Re-evaluating Foucault's Legacy. In: Organization 14(6): 769-791.

Caravani, Alice, Neil Bird, et al. (2011). REDD-plus Finance. Climate Finance Fundamentals, Brief 5. Washington D.C.: Heinrich Böll Stiftung/Overseas Development Institute.

Cerny, Philip (1990). The Changing Architecture of Politics: Structure, Agency, and the Future of the State London: Sage.

Cerny, Philip (2008). Embedding Neoliberalism: The Evolution of a Hegemonic Paradigm. In: The Journal of International Trade and Diplomacy 2(1): 1-46.

Cerny, Philip and Mark Evans (1999). New Labour, Globalization, and the Competition State. Centerfor European Studies Working Paper Series #70

Chasek, Pamela, (Ed.) (2010). Earth Negotiations Bulletin. Summary of the Bonn Climate Change Talks: 31 May - 11 June 2010. Bonn: International Institute for Sustainable Development.

Chasek, Pamela, (Ed.) (2010). Summary of the Cancun Climate Change Conference: 29 November - 11 December 2010. New York: IISD.

Clifton, Sarah-Jayne (2009). A dangerous obsession. The evidence against carbon trading and for real solutions to avoid a climate crunch. London: Friends of the Earth.

Climate Focus (2011). Briefing Note: CP16/CMP 6: The Cancún Agreements. January 2011. Amsterdam: Climate Focus.

Climate Policy Initiative (2011). Review of Low Carbon Development in China: 2010 Report. Bejing.

Cline, William (1992). The Economics of global warming. Washington: Institute for International Economics.

CNA (2007). National Security and the thread of climate change. Alexandria/Virginia: The CNA Corporation.

Collier, Stephen J. (2009). Topologies of Power. Foucault's Analysis of Political Government beyond 'Governmentality'. In: Theory, Culture & Society 26(6): 78-108.

Corbin, Juliet and Anselm Strauss (1990). Grounded Theory Research: Procedures, Canons, and Evaluative Criteria. In: Qualitative Sociology 13(1): 3-21.

Costanza, Robert, Ralph d'Arge, et al. (1997). The value of the world's ecosystem services and natural capital. In: Nature 387: 253-260.

Cozijnsen, Jos, Daniel Dudek, et al. (2007). CDM and the Post-2012 Framework. Wien: Environmental Defense.

Crouch, Colin (2011). The Strange Non-Death of Neoliberalism. Cambridge: Polity Press.

Daly, Herman (1998). The return of Lauderdale's paradox. In: Ecological Economics 25: 21-23.

de Goede, Marieke (2005). Virtue, Fortune and Faith: A Genealogy of Finance. Minneapolis: University of Minnesota Press.

Dean, Mitchell (1994). Critical and Effective Histories. Foucault's Methods and Historical Sociology. London/New York: Routledge.

Dean, Mitchell (2010). Governmentality. Power and rule in modern society. Sage.

Death, Carl (2011). Summit theatre: exemplary governmentality and environmental diplomacy in Johannesburg and Copenhagen. In: Environmental Politics 20(1): 1-19.

Deeg, Richard and Mary O'Sullivan (2009). Review Article: The political economy of Global finance capital. In: World Politics 61(4): 731–763.

Deleuze, Gill (1992). What is a dispositif, in: Thimothy Armstrong: Michel Foucault: Philosopher. Hemel Hempstead: Harvester Wheatsheaf, 159-16

Denninger, Tina, Silke van Dyk, et al. (2010). Die Regierungs des Alter(ns). Analysen im Spannungsfeld von Diskurs, Dispositiv und Disposition, in: Johannes Angermüller and Silke van Dyk: Diskursanalyse meets Gouvernementalitätsforschung. Zur Einführung. Frankfurt am Main: Campus, 207-235.

Descheneau, Philippe and Matthew Paterson (2011). Between Desire and Routine: Assembling Environment and Finance in Carbon Markets. In: Antipode 43(3): 662–681.

Deutscher Bundestag (2009). Antwort der Bundesregierung auf die Kleine Anfrage der Abgeordneten Dr. Gerhard Schick, Alexander Bonde, Kerstin Andreae, weiterer Abgeordneter und der Fraktion BUNDNIS 90/DIE GRUNEN: Nachhaltigkeit bei der Anlagestrategie der öffentlichen Hand. Drucksache 16/12018. Berlin: Deutscher Bundestag.

Diaz-Bone, Rainer and Getraude Krell, (Eds.) (2009). Diskurs und Ökonomie. Diskuranalytische Perspektiven auf Märkte und Organisationen. VS Verlag: Wiesbaden.

Dicken, Peter (2003). Global Shift: Reshaping the Global Economic Map in the 21st Century. London: The Guilford Press.

Diefenbach, Katja (1999). Kapitalismus verstehen. Poststrukturalistische Mikropolitiken bei Guattari, Deleuze und Foucault, in: jour-fixe-initiative berlin: Kritische Theorie und Poststrukturalismus. Theoretische Lockerungsübungen. Berlin/ Hamburg: Argument, 79-95.

Dietz, Kristina (2006). Vulnerabilität und Anpassung gegenüber Klimawandel aus sozial-ökologischer Perspektive. Diskussionspapier 01/06 des Projektes „Global Governance und Klimawandel". Berlin.

Doornbosch, Richard and Eric Knight (2008). What Role for public finance in international climate change mitigation. Paris: OECD.

Dowling, Robyn (2010). Geographies of identity: climate change, governmentality and activism. In: Progress in Human Geography 34(4): 488-495.

Dreyfus, Hubert and Paul Rabinow (1982). Michel Foucault. Beyond Structuralism and Hermeneutics. Chicago: University of Chicago Press.

du Gay, Paul and Michael Pryke (2002). Cultural Economy: An introduction, in: Dies.: Cultural Economy. London: Sage, 1-20.

Dutschke, Michael and Sheila Wertz-Kanounnikoff (2008). Financing REDD. Linking country needs and financing sources: CIFOR Infobrief No.17. http://www.cifor.cgiar.org/publications/pdf_files/Infobrief/017-infobrief. pdf. Retrieved: 20.03.2012.

Eckersley, Robyn (2000). Disciplining the market, calling in the state: the politics of economy-environment integration, in: S. Young: The Emergence of Ecological Modernisation: Integration the Environment and the Economy. London: Routledge, 233-252.

Eckersley, Robyn (2004). The green state: rethinking democracy and souvereignty. Cambridge/London: The MIT Press.

Egner, Heike (2007). Überraschender Zufall oder gelungene wissenschaftliche Kommunikation: Wie kam der Klimawandel in die aktuelle Debatte? In: GAIA 16(4): 250-254.

EIB (2009). EIB Launches Debut Swedish Krona Climate Awareness Bonds. http://www.eib.org/investor_relations/press/2009/2009-215-eib-launches-debut swedish-krona-climate-awareness-bonds.htm?lang=en

Eliasch, John (2009). Climate change: financing global forests. The Eliasch Review. London: Earthscan.

Ellis, Jane, Jan Corfee-Morlot, et al. (2004). Taking stock of progress unter the clean development mechanism. Paris: International Energy Agency.

Epstein, Gerald (2005). Introduction: Financialization and the World Economy, in: Gerald Epstein: Financialization and the World Economy. Cheltenham: Edward Elgar, 3-16.

Ernst&Young (2007). Strategic Business Risk 2008 – the Top 10 Risks for Business. London. Retrieved:

Euractiv (2009). "World Bank: 'Offsets will play a role post-Kyoto'. Interview with Joëlle Chassard, Manager of the carbon finance unit at the World Bank." Retrieved: 05.06.2009, from: http://www.euractiv.com/en/climate-change/world-bank-offsets-play-role-post-kyoto/article-182486.

European Commission (2007). Adapting to climate change in Europe – options for EU action. Green Paper of the European Commission. COM(2007) 354 final. Brussels.

European Commission (2008). Proposal for a decision of the European Parliament and of the council on the effort of Member States to reduce their greenhouse gas emissions to meet the Community´s greenhouse gas reduction commitments up to 2020. Brussels.

European Commission (2008). The support of electricity from renewable energy sources. Commission staff working document. SEC(2008)57. Brussels.

European Commission (2011). A Roadmap for moving to a competitive low carbon economy in 2050. Brussels.

Ewald, Francois (1978). Foucault - Ein vagabundierendes Denken, in: Michel Foucault: Dispositive der Macht. Über Sexualität, Wissen und Wahrheit. Berlin: Merve, 7-20.

Ewald, Francois (1991). Insurance and Risk, in: Graham Burchell, Colin Gordon and Peter Miller: The Foucault Effect. Studies in Governmentality. Chicago: The University of Chicago Press, 197-210.

Falkner, Robert (2010). Business and global climate governance: a neo-pluralist perspective, in: Morten Ougaard and Anna Leander: Business and global governance. London: Routledge.

Fankhauser, Sam (2006). "The Economics of Adaptation." Retrieved: 20.05.2009, from: www.hm-treasury.gov.uk/d/stern_review_supporting _technical_material_sam_fankhauser_231006.pdf.

Flavin, Christopher (2009). Low-Carbon Energy: A roadmap. Washington D.C.: Worldwatch Institute.

Flavin, Christopher (2009). Low-Carbon Energy. A Roadmap. Washington D.C.: Worldwatch Institute.

Fligstein, Neil (2001). The Architecture of markets. An Economic sociology of twenty-first-century capitalist societies. Princeton: Princeton University Press.

Foucault, Michel (1973). The Order of Things: An Archaeology of the Human Sciences. New York: Vintage/Random House.

Foucault, Michel (1977). Nietzsche, Genealogy, History, in: D. F. Bouchard: Language, Counter-Memory, Practice: Selected Essays and Interviews. Ithaca: Cornell University Press, 139-164.

Foucault, Michel (1978b). Ein Spiel um die Psychoanalyse. Gespräch mit Angehörigen des Department de Psychanalye der Universität Paris/ Vincennes, in: Dispositive der Macht. Michel Foucault über Sexualität, Wissen und Wahrheit. Berlin: Merve, 118-175.

Foucault, Michel (1978c). Die Machtverhältnisse durchziehen das Körperinnere. Gespräch mit Lucette Finas, in: Dispositive der Macht. Michel Foucault über Sexualität, Wissen und Wahrheit. Berlin: Merve, 104-117.

Foucault, Michel (1980a). The Confession of the Flesh. Interview, in: Colin Gordon: Power/Knowledge. Selected Interviews and Other Writings. New York: Pantheon Books, 194-228.

Foucault, Michel (1980b). Truth and Power, in: Colin Gordon: Power/Knowledge. Selected Interviews and other writings 1972-1977. New York: Pantheon Books, 109-133.

Foucault, Michel (1981). History of Systems of Thought 1979. In: Philosophy & Social Criticism 8: 353-359.

Foucault, Michel (1982). The Subject and Power, in: Hubert Dreyfus and Paul Rabinow: Michel Foucault: Beyond Structuralism and Hermeneutics. London: Harvester Wheatsheaf, 208-226.

Foucault, Michel (1986). The Use of Pleasure: The History of Sexuality: Volume Two New York: Vintage.

261

Foucault, Michel (1989b). The Archaeology of Knowledge. London, New York: Routledge.

Foucault, Michel (1991). Questions of method, in: Graham Burchell, Colin Gordon and Peter Miller: The Foucault Effect. Studies in Governmentality. Chicago: The University of Chicago Press.

Foucault, Michel (1998). The History of Sexuality Vol. 1: The Will to Knowledge. London: Penguin.

Foucault, Michel (2001). Fearless Speech. New York: Semiotext.

Foucault, Michel (2005). Polemics, Politics and Problematization. Interview with P. Rabinow, May 1984, in: Paul Rabinow: Ethics: Subjectivity and Truth. Essential Works of Foucault, 1954-1984. New York: New Press, 111-120.

Foucault, Michel (2007). Security, Territory, Population. Lectures at the College de France. Basingstoke: Palgrave Macmillan.

Foucault, Michel (2008). The Birth of Biopolitics. Lectures at the College de France. New York: Palgrave Macmillan.

Fougner, Tore (2006). The state, international competitiveness and neoliberal globalisation: is there a future beyond 'the competition state'? . In: Review of International Studies 32: 165–185.

Friedman, Thomas (2007). The Power of Green: New York Times Magazine. http://www.nytimes.com/2007/04/15/magazine/15green.t.html?pagewanted =all. Retrieved: 20.03.2012.

Fuhr, Harald and Markus Lederer (2009). Varieties of Carbon Governance in Newly Industrializing Countries. In: The Journal of Environment Development 18(4): 327-345.

GIB (2010). Unlocking investment to deliver Britain's low carbon future. London: Green Investment Bank Commission.

Giddens, Anthony (2009). The Politics of Climate Change. Cambridge: Polity Press.

Gomez-Echeverri, Luis and Benito Müller (2009). The financial mechanism of the UNFCCC. A brief history. Oxford: European capacity building initiative.

Gordon, Colin (1991). Governmental Rationality: An introduction, in: Graham Burchell, Colin Gordon and Peter Miller: The Foucault Effect. Studies in Governmentality. Chicago: University of Chicago Press, 1-51.

Gordon, Leonie, (Ed.) (2011). World Future Energy Summit Bulletin. A Summary Report of the World Future Energy Summit (WFES) 2011. Winnipeg: IISD.

Gore, Tim (2010). Climate Finance Post-Copenhagen. Brussels: Oxfam International.

Green Korea United (2009). Green Growth Policy of the Korean Government and Its Critics: Korea NGO Network. http://green-korea.tistory.com/101. Retrieved: 03.07.2011.

Green New Deal Group (2008). A Green New Deal. Joined-up policies to solve the triple crunch of the credit crisis, climate change and high oil prices. London: New Economics Foundation.

Gupta, Joyeeta (2008). Post 2012: CDMs Role in the Climate Negotiations. Policy Note for the European Parliament 08-26. Amsterdam: Institute for Environmental Studies.

Gutman, Pablo and Bruce Cabarle (2010). At the crossroads? The nexus between public and private finance for REDD+. Draft paper. Washington D.C.: WWF.

Hajer, Marteen (1997). The Politics of Environmental Discourse: ecological modernization and the policy process. London: Oxford University Press.

Hamilton, Kirsty (2008). Clean Energy Investment and the 'New Competitiveness'. London: Chatham House.

Hamilton, Kirsty (2009). Unlocking Finance for Clean Energy: The Need for 'Investment Grade' Policy. London: Chatham House.

Hanson, Norwood Russel (1972). Observation and Explanation. London: George Allen and Unwin.

Hart, Keith and Chris Hann (2009). Introduction: Learning from Polanyi 1, in: Keith Hart and Chris Hann: Market and Society. The Great Transformation Today. Cambridge: Cambridge University Press, 1-16.

Helm, Dieter (2009). EU climate-change policy - a critique, in: Dieter Helm and Cameron Hepburn: The Economics and Politics of Climate Change. Oxford: Oxford University Press, 222-244.

Herendeen, Robert (1998). Monetary-costing environmental services: nothing is lost, something is gained. In: Ecological Economics 25: 29-30.

Hertsgaard, Mark (1998). Earth Odyssey. Around the World in Search for our Environmental Future. New York: Broadway Books.

Hertsgaard, Mark (2001). "A green new deal." Retrieved: 28.03.2011, from: http://motherjones.com/politics/2001/06/new-green-deal.

Hindess, Barry (1996). Discourses of Power. From Hobbes to Foucault. Oxford: Blackwell.

Hirsch, Joachim (1995). Der national Wettbewerbsstaat. Berlin: Edition ID-Archiv.

Hohnen, Paul (1999). Greenpeace & the financial sector - the possibility of profitable relationships between non-for-profits and for-profits. Berlin: Global Public Policy Institute.

Holm Olsen, Karen (2007). The clean development mechanism's contribution to sustainable development: a review of the literature. In: Climatic Change 87: 59-73.

Honneth, Axel (1984). Kritik der Macht. Reflexionsstufen einer kritischen Gesellschaftstheorie. Frankfurt am Main: Suhrkamp.

Honneth, Axel and Martin Saar (2008). Nachwort: Geschichte der Gegenwart. Michel Foucaults Philosophie der Kritik, in: Michel Foucault. Die Hauptwerke. Frankfurt am Main: Suhrkamp, 1651-1682.

Howarth, David (2000). Discourse. Buckingham: Open University Press.

Howarth, David and Jason Glynos (2007). Logics of Critical Explanation in Social and Political Theory. London: Routledge.

HSBC (2010). Climate Confidence Monitor 2010. London: HSBC.

Hughes, Hannah (2011). The practice of writing and the Intergovernmental Panel on Climate Change (IPCC). Paper presented at the 6th International Conference in Interpretive Policy Analysis: Discursive Spaces, Politics, Practices and Power, Cardiff, June 2011.

Hulme, Mike (2007b). The Limits of the Stern Review for Climate Change Policy-making. http://mikehulme.org/wp-content/uploads/2007/04/2007 _03-british-ecological-society.pdf. Retrieved: 19.12.2011.

Hulme, Mike (2010). After the crash - a new direction for climate policy. BBC News Viewpoint. http://news.bbc.co.uk/2/hi/science/nature/8673828.stm. Retrieved: 19.12.2011.

Humphrey, John (2004). The Clean Development Mechanism: how to increase benefits for developing countries. In: IDS Bulletin 35(3): 84–89.

IIGCC et al. (2009). 2009 Investor Statement on the Urgent Need for a Global Agreement on Climate Change. Geneva: UNEP Finance Initiative.

IIGCC et al. (2010). Global Investor Statement on Climate Change: Reducing Risks, Seizing Opportunities & Closing the Climate Investment Gap. Geneva: UNEP Finance Initiative.

IWG IFR (2009). "Report of the Informal Working Group on Interim Finance for REDD+ (IWG IFR), October 2009." from: http://www.unredd. net/index.php?option=com_docman&task=doc_details&Itemid=&gid=109.

Jacob, Daniela et al. (2008). Klimaauswirkungen und Anpassung in Deutschland – Phase 1: Erstellung regionaler Klimaszenarien für Deutschland. Dessau: Umweltbundesamt.

Jänicke, Martin (2008). Ecological modernisation - new perspectives. In: Journal of Cleaner Production 16(5): 557-565.

Jessop, Bob (2005). Macht und Strategie bei Poulantzas und Foucault. Supplement der Zeitschrift Sozialismus. Hamburg: VSA.

Jessop, Bob (2007). From micro-power to governmentality: Foucault´s work on statehood, state formation, statecraft and state power. In: Political Geography 26: 34-40.

Jessop, Bob and Stijn Oosterlynck (2008). Cultural Political Economy: On making the cultural turn without falling into soft economic sociology. In: Geoforum 39(3): 1155-1169.

Johnson, Todd , Claudio Alatorre, et al. (2010). Low-carbon development for Mexico. Washington D.C.: World Bank.

Jones, Randall and Byungseo Yoo (2010). Korea's Green Growth Strategy: Mitigating Climate Change and Developing New Growth Engines. Paris: OECD Publishing.

Justice, Sophie (2009). Private financing of Renewable Energy. A guide for policymakers. London: Chatham House.

Karousakis, Katia and Jan Corfee-Morlot (2007). Financing mechanisms to reduce emissions from deforestation: Issues in design and implementation. Paris: OECD/ IEA.

Kelly, Mark (2009). The political philosophy of Michel Foucault. New York/London: Routledge.

Keohane, Robert and David Victor (2010). The regime complex for climate change. Harvard: Harvard Kennedy School.

Kerchner, Brigitte (2006). Wirklich Gegendenken. Diskursanalyse mit Michel Foucault, in: Brigitte Kerchner and Silke Schneider: Foucault: Diskusanalyse der Politik. Wiesbaden: VS Verlag für Sozialwissenschaften, 145-164.

Keskitalo, Carina, Sirkku Juhola, et al. (2012). Climate change as governmentality: technologies of government for adaptation in three European countries. In: Journal of Environmental Planning and Management 55(4): 435-452.

Koch-Weser, Caio (2011). Durban must clear the path for more low carbon investment. Natural Resources Policy Paper. London: Deutsche Bank Research.

Laclau, Ernesto (2000). Identity and Hegemony, in: Judith Butler, Ernesto Laclau and Slavoj Žižek: Contingency, Hegemony, Universality. London, New York: Verso, 44-89.

Laclau, Ernesto and Chantal Mouffe (1985). Hegemony and Socialist Strategy. London: Verso.

Lall, Sanjaya (2001). Competitive Indices and Developing Countries: An Economic Evaluation of the Global Competitiveness Report. In: World Development 29(9): 1501-1525.

Landell-Mills, Natasha (2002). Developing markets for forest environmental services: an opportunity for promoting equity while securing efficiency? In: Philosophical transactions of the royal society 360: 1817-1825.

Lansing, David (2011). Realizing Carbon's Value: Discourse and Calculation in the Production of Carbon Forestry Offsets in Costa Rica. In: Antipode 43(3): 731-753.

Leal Riesco, Iola and Kyeretwie Opoku (2009). Is REDD undermining FLEGT? Avoiding Deforestation and Degradation Briefing Note 5. Moreton-in-Marsh: FERN.

Legget, Jeremy (1993). Climate Change and the Insurance Industry. Solidarity among the Risk Community?: Greenpeace.

Legget, Jeremy (2001). The carbon war. Global warming and the end of the oil area. New York: Routledge.

Lemke, Thomas (1997). Eine Kritik der politischen Vernunft. Foucaults Analyse der modernen Gouvernmentalität. Hamburg: Argument.

Lemke, Thomas (2002). Foucault, Governmentality, and Critique. In: Rethinking Marxism 14(3): 49-64.

Lemke, Thomas (2007). Gouvernementalität und Biopolitik. Wiesbaden: VS Verlag.

Levy, David and Daniel Egan (2003). A Neo-Gramscian Approach to Corporate Political Strategy: Conflict and Accommodation in the Climate Change Negotiations. In: Journal of Management Studies 40(4): 803-829.

Lohmann, Larry (2009). Climate as Investment. In: Development and Change 40(6): 1063-1083.

Lövbrand, Eva and Johannes Stripple (2011). Making climate change governable: Accounting for sinks, credits and personal budgets. In: Critical Policy Studies 5(2): 187-200.

Lövbrand, Eva, Johannes Stripple, et al. (2009). Earth System Governmentality: Reflections on Science in the Anthropocene. In: Global Environmental Change 19: 7-13.

Lovell, Heather and Diana Liverman (2010). Understanding carbon offset technologies. In: New Political Economy 15(2): 255-273.

Lovell, Heather and Donald MacKenzie (2011). Accounting for Carbon: The Role of Accounting Professional Organisations in Governing Climate Change. In: Antipode 43(3): 704-730.

Lubowski, Ruben (2008). The role of REDD in stabilising greenhouse gas concentrations. Lessons from economic models. Bogota: CIFOR.

Luhmann, Niklas (1995). Social Systems. Stanford: Stanford University Press.

Luke, Timothy (1996). Generating Green Governmentality: A Cultural Critique of Environmental Studies as a Power/Knowledge Formation. Virginia Polytechnic Institute. http://www.cddc.vt.edu/tim/tims/Tim514a.PDF. Retrieved: 20.03.2012.

Luke, Timothy (1999a). Environmentality as Green Governmentality, in: E. Darier: Discourses of the Environment. Oxford: Blackwell Publishers, 121-151.

Luke, Timothy (2009). A green new deal: why green, how new, and what is the deal? In: Critical Policy Studies 3(1): 14-28.

Luks, Fred (2008). Der Diskurs über das Klima und das Klima des Diskurses. In: GAIA 17(2): 186-188.

MacKenzie, D. and Y. Millo (2003). Constructing a Market, Performing Theory: The Historical Sociology of a Financial Derivatives Exchange. In: American Journal of Sociology 109(1): 107–145.

MacKenzie, Donald (2009). Making Things the Same: Gases, Emission Rights and the Politics of Carbon Markets. In: Accounting, Organizations and Society 34: 440-455.

Maclean, John, Jason Tan, et al. (2008). Public Finance Mechanisms to Mobilise Investment in Climate Change Mitigation. Paris: UNEP Sustainable Energy Finance Initiative.

MacLeod, Michael (2010). Private Governance and Climate Change: Institutional Investors and Emerging Investor-Driven Governance Mechanisms. In: St Antony's International Review 5(2): 46-65.

MacNeil, Robert and Matthew Paterson (2012). Neoliberal climate policy: from market fetishism to the development state. In: Environmental Politics 21(2): 230-247.

Mandelson, Peter (2009). Climate change: the political and business challenge, in: Anthony Giddens, Simon Latham and Roger Little: Building a low-carbon future: The politics of climate change. London: Policy Network, 89-98.

Maréchal, K (2007). The economics of climate change and the change of climate in economics. In: Energy Policy 35: 5181-5194.

Martone, Francesco (2010). The emergence of the REDD hydra. An analysis of the REDD-related discussions and developments in the June session of the UNFCCC and beyond. Moreton-in-Marsh: Forest Peoples Programme.

Mc Kinsey (2009). Pathways to a low carbon Economy. Executive Summary. London: Mc Kinsey.

Meckling, Jonas (2011). Carbon Coalitions. Business, Climate Politics, and the Rise of Emissions Trading. Massachusets: Massachusets Instiute of Technology.

Methmann, Chris and Delf Rothe (2011). Politics in the Day after Tomorrow: The Political Effect of Apocalyptic Imaginaries in Global Climate Governance. Paper for the International Studies Association Annual Conference "Global Governance: Political Authority in Transition", MONTREAL, QUEBEC, CANADA, Mar 16, 2011.

Michaelowa, Axel and Katharina Michaelowa (2007). Does climate policy promote development? In: Climatic Change 84: 1-4.

MIGA (2010). Bolstering Private Equity Investment. MIGA fact sheet. Washington D.C.: World Bank.

Miller, Peter (2008). Calculating economic life. In: Journal of Cultural Economy 1(1): 51-64.

Mills, Evan (2009). From Risk to Opportunity. Insurer Responses to Climate Change. Boston: Ceres.

Moebius, Stephan and Andreas Reckwitz (2008). Poststrukturalismus und Sozialwissenschaften. Eine Standortbestimmung, in: Stephan Moebius and Andreas Reckwitz: Poststrukturalistische Sozialwissenschaften. Frankfurt am Main: Suhrkamp, 7-23.

Möhner, Annett and Richard Klein (2007). The Global Environment Facility: Funding for Adaptation or Adapting to Funds? Stockholm: Stockholm Environment Institute.

Mostert, Wolfgang (2010). Publicly based guarantees as policy instruments to promote clean energy. Basel: UNEP SEF Alliance.

Muller, Adrian (2007). How to make the clean development mechanism sustainable. The potential of rent extraction. In: Energy Policy 35: 3203-3212.

Müller, Benito (2009a). Are treasuries killing the climate deal? Oxford: Oxford Institute for Energy Studies.

Müller, Benito (2009c). 'Under the Authority of the COP'? Oxford Energy and Environment Comment, November 2009. Oxford: Oxford Institute for Energy Studies.

Newell, Peter, Nicky Jenner, et al. (2009). Governing Clean Development. A framework for analysis. Norwich: University of East Anglia.

Newell, Peter and Matthew Paterson (2010). Climate Capitalism. Cambridge: Cambridge University Press.

Nölke, Andreas and James Perry (2007). The Power of Transnational Private Governance: Financialization and the IASB. In: Business and Politics 9(3).

Nordhaus, William (2007). A Question of Balance. Weighing the Options on Global Warming Policies. New Haven/London: Yale University Press.

Norgaard, Richard and Collin Bode (1998). Next, the value of god, and other reactions. In: Ecological Economics 25: 37-39.

O'Brien, Karen L. and Robin Leichenko (2000). Double exposure: assessing the impacts of climate change within the context of economic globalization. In: Global Environmental Change 10: 221-232.

Oberthür, Sebastian and Hermann Ott (1999). Das Kyoto-Protokoll. Klima-Politik für das 21. Jahrhundert.

Oberthür, Sebastian and Hermann Ott (1999). The Kyoto Protocol. International Climate Policy for the 21st Century. Berlin: Springer.

Oels, Angela (2005). Rendering climate change governable: From biopower to advanced liberal government. In: Journal of Environmental Policy and Planning 7(3): 185-208.

Oels, Angela (2012). From 'securitization' of climate change to 'climatization' of the security field: Comparing three theoretical perspectives, in: Jürgen Scheffran, Michael Broszka, Hans-Günter Brauch, Michael Link and Janpeter Schilling: Climate Change, Human Security and Violent Conflict. Hexagon Series, Vol. VIII New York: Springer, 185-205.

Onstwedder, Jan-Peter and Michael Mainelli (2010). Living Up To Their Promises (index-linked carbon bonds). In: Environmental Finance(1 February 2010): 17.

Owen, David (1995). Genealogy as exemplary critique: reflections on Foucault and the imagination of the political. In: Economy and Society 24(4): 489-506.

Palan, Ronen and Jason Abbot (1996). State Strategies in Global Political Economy. London: Pinter.

Palley, Thomas (2007). Financialization: What It Is and Why It Matters. Paper presented at the conference "Finance-led Capitalism? Macroeconomic Effects of Changes in the Financial Sector," Berlin, Germany, October 26–27, 2007. The Levy Economics Institute.

Parker, Charlie (2011). Policy Brief: The Outcome for Forests Emerging from Cancun. http://www.globalcanopy.org/updates/blogs/policy-brief-outcome-forests-emerging-cancun. Retrieved: 5 September 2011

Parker, Charlie, Jessica Brown, et al. (2009). The little climate finance book. Oxford: Global Canopy Programme.

Parker, Charlie, Andrew Mitchell, et al. (2009). The little REDD+ book. Oxford: Global Canopy Programme.

Paterson, Matthew (2010). Legitimation and Accumulation in Climate Change Governance. In: New Political Economy, 15(3): 345-368.

Paterson, Matthew and Johannes Stripple (2010). My Space: Governing Individuals' Carbon Emissions. In: Environment and Planning D: Society and Space 28: 2341-2362.

Pattberg, Philipp and Johannes Stripple (2008). Beyond the public and private divide: remapping transnational climate governance in the 21st century. In: International Environmental Agreements(8): 367–388.

Pearce, David (2003). The social cost of carbon and its policy imlications. In: Oxford Review of Economic Policy 19(3): 362-384.

Pearce, Fred (1997). "Dear Greenpeace." New Scientist, 15.11.1997. Pages: 54

Pearson, Ben (2007). Market failure: why the Clean Development Mechanism won't promote clean development. In: Journal of Cleaner Production 15: 247-252.

Pendleton, Andrew and Simon Retallack (2009). Fairness in Global Climate Change Finance. London: Institute for Public Policy Research.

Peskett, Leo, David Huberman, et al. (2008). Making REDD work for the poor. London: Overseas Development Institute.

Phelps, Jacob, Edward Webb, et al. (2010). Does REDD+ Threaten to Recentralize Forest Governance? In: Science 328: 312-313.

Phillips, Kevin (2006). American Theocracy: The Peril and Politics of Radical Religion, Oil, and Borrowed Money in the 21st Century. London: Penguin Books.

Pielke, Roger, Gwyn Prins, et al. (2007). Climate change 2007: Lifting the taboo on adaptation. In: Nature 445: 597–598

Policy Network (2009). The politics of climate change: Our role in the debate, in: Anthony Giddens, Simon Latham and Roger Little: Building a low-carbon future: The politics of climate change. London: Policy Network, 12-19.

Porter, Gareth, Neil Bird, et al. (2008). New Finance for Climate Change and the Environment. Washington DC: WWF/Heinrich Böll Foundation.

Power, Michael (2000). The Audit Society – Second Thoughts. In: International Journal of Auditing 4(1): 111-119.

Presidential Committee on Green Growth (2009). National Strategy for Green Growth and Five-Year Plan. Seoul: Republic of Korea.

Price, Colin (1993). Time, Discounting and Value. Oxford: Basil Blackwell.

Prins, Gwyn, Isabel Galiana, et al. (2010). The Hartwell Paper. A new direction for climate policy after the crash of 2009. London School of Economics, University of Oxford.

Prins, Gwyn and Steve Rayner (2007a). Time to ditch Kyoto. In: Nature 449: 973-975.

Prins, Gwyn and Steve Rayner (2007b). The Wrong Trousers: Radically Rethinking Climate Policy. Discussion Paper. University of Oxford, London School of Economics.

Project Catalyst (2009a). Scaling up climate finance. Finance briefing paper. Brüssel: European Climate Foundation, Climate Works.

Project Catalyst (2009b). Financing Global Action on Climate Change. Brüssel: European Climate Foundation, Climate Works.

Raworth, Kate (2007). Adaptation to Climate Change. What's needed in poor countries and who should pay. Oxford: Oxfam International.

Rayner, Steve (2010). How to eat an elephant: A bottom-up approach to climate policy. In: Climate Policy 10(6): 615-621.

Reed, David, Pablo Gutman, et al. (2009). New Mechanisms for Financing Mitigation. Transforming economies sector by sector. Washington D.C.: WWF.

Republic of Guyana - Office of the President (2010). Transforming Guyana's Economy While Combating Climate Change. A Low-Carbon Development Strategy.

Robertson, Morgan (2004). The neoliberalization of ecosystem services: wetland mitigation banking and problems in environmental governance. In: Geoforum 35: 361-373.

Robins, Nick and Mark Fulton (2009). Investment Opportunities and Catalysts. Analysis and proposals from the Climate Finance Industry on Funding Climate Mitigation, in: Richard Stewart, Benedict Kingsbury and Bryce Rudyk: Climate Finance. Regulatory and Funding Strategies for Climate Change and Global Development. New York/London: New York University Press, 143-151.

Romani, Mattia (2009). Meeting the Climate Challenge: Using Public Funds to Leverage Private Investment in Developing Countries. Summary for policy makers. London: London School of Economics.

Rose, Nikolas, Pat O'Malley, et al. (2006). Governmentality. In: Annual Review of Law and Society 2(83–104).

Rothe, Delf (2011). Cleaning Foucault's Glasses. Problems and blind-spots of a governmen- tality approach to global climate governance. Paper prepared for the workshop: Governing the global climate polity: Rationality, practice and power, Lund University, Sweden, 19-21 June 2011.

Rotmans, J. and H. Dowlatabadi (1998). Integrated assessment modeling, in: Steven Rayner and Elizabeth Malone: Human choice and climate change. Vol 3: the tools for policy analysis. Columbus: Battelle Press, 291–377.

Rubio Alvarado, Laura Ximena and Sheila Wertz-Kanounnikoff (2007). Why are we seing "REDD"? An anlysis of the international debate on reducing emissions from deforestation and degradation in developing countries. Paris: IDDRI.

Rutherford, Stephanie (2007). Green governmentality: insights and opportunities in the study of nature's rule. In: Progress in Human Geography 31(3): 291-307.

Schalatek, Liane (2009). Gender and Climate Finance. Washington D.C.: Heinrich Böll Foundation.

Schipper, Lisa (2006). Conceptual History of Adaptation in the UNFCCC Process. In: Review of European Community & International Environmental Law 15(1): 82-92.

Schipper, Lisa and Ian Burton (2009). Understanding adaptation: Origins, concepts, practice and policy, in: Lisa Schipper and Ian Burton: The Earthscan Reader on Adaptation to Climate Change. Oxford: Earthscan, 1-8.

Schivelbusch, W. (2007). Three new deals: reflections of Roosevelt's America, Mussolini's Italy, and Hitler's Germany 1933-1939. New York: Picador.

Schneider, Lambert (2007). Is the CDM fulfilling its environmental and sustainable development objectives? An evaluation of the CDM and options for improvement. Freiburg: Öko-Institut.

Schneider, Stephen (2008). The Stern Review debate: an editorial essay. In: Climatic Change 89: 241-244.

Scholz, Imme and Lars Schmidt (2008). Reducing Emissions from Deforestation and Forest Degradation in Developing Countries: Meeting the main challenges ahead. DIE Briefing Paper 6/2008. Bonn: German Development Institute.

Scholz, Imme and Laura Schmidt (2009). Financing the climate agenda: the development perspective. Bonn: DIE-GDI/BMZ.

Sierra, Katherine (2007). Commentary: G-8 Summit and Climate Change: The World Bank. http://climatechange.worldbank.org/news/g-8-summit-and-climate-change. Retrieved: 08.03.2011.

Sierra, Katherine (2011). Message for Durban: Scale-up through a Green Climate Fund Private Sector Facility: The Brookings Institution. http://www.brookings.edu/opinions/2011/1128_durban_sierra.aspx?p=1. Retrieved: 08.12.2011.

Smit, Barry and Olga Pilifosova (2001). Adaptation to Climate Change in the Context of Sustainable Development and Equity, in: James J. et al. McCarthy: IPCC Third Assessment Report: Climate Change 2001: Working Group II: Impacts, Adaptation and Vulnerability. Cambridge: Cambridge University Press, 879-912.

Spash, Clive (2007). The economics of climate change impacts à la Stern: Novel and nuanced or rhetorically restricted? In: Ecological Economics 63: 706-713.

Spash, Clive (2010). The brave new world of carbon trading. In: New Political Economy 15(3): 169-195.

Spence, Chris, (Ed.) (2009). Briefing Note on Financing the Climate Agenda. IISD Reporting Service. New York: International Institute for Sustainable Development.

Stäheli, Urs (2008a). Ökonomie. Die Grenzen des Ökonomischen, in: Stephan Moebius and Andreas Reckwitz: Poststrukturalistische Sozialwissenschaften. Frankfurt am Main: Suhrkamp, 295-311.

Steer, Andrew (2011). Will Durban deliver? Blog Post by Andrew Steer, World Bank Special Envoy for Climate Change. http://blogs. worldbank.org/ climatechange/will-durban-deliver. Retrieved: 03.12.2011.

Stephan, Benjamin (2012). Bringing discourse to the market: commodifying avoided deforestation. In: Environmental Politics 21(4): 621-639.

Stern, Nicholas (2006). The Economics of Climate Chance. The Stern Review Cambridge: Cambridge University Press.

Stern, Nicholas (2008). Key Elements of a global deal on climate change. London: London School of Economics.

Stern, Nicholas (2009a). A blueprint for a Safer Planet. New York: Random House.

Stern, Nicholas (2009b). The Global Deal: Climate Change and the Creation of a New Era of Progress and Prosperity. New York: Public Affairs.

Stewart, Richard, Benedict Kingsbury, et al. (2009a). Climate Finance for Limiting Emissions and Promoting Green Development: Mechanisms, Regulation, and Governance, in: ibid.: Climate Finance. Regulatory and Funding Strategies for Climate Change and Global Development. New York: New York University Press, 3-34.

Streck, Charlotte (2009a). Expectations and Reality of the Clean Development Mechanism: A Climate Finance Instrument between Accusation and Aspirations, in: Richard Stewart, Benedict Kingsbury and Bryce Rudyk: Climate Finance. Regulatory and Funding Strategies for Climate Change and Global Development. New York/London: New York University Press, 67-75.

Streck, Charlotte (2009b). Financing REDD. A review of selected policy proposals. Washington D.C.: WWF.

Stripple, Johannes (2010). Weberian Climate Policy: Administrative rationality organized as a market, in: Karin Bäckstrand, J. Khan, A. Kronsell and Eva Lövbrand: Environmental Politics and Deliberative Democracy. Cheltenham: Edward Elgar Publishing, 67-84.

Suárez Müller, Fernando (2004). Skepsis und Geschichte. Das Werk Michel Foucaults im Lichte des absoluten Idealismus. Würzburg: Königshausen und Neumann.

Sukhdev, Pavan (2009). Global Green New Deal. An Update for the G20 Pittsburgh Summit: United Nations Environment Programme. http://www. unep.org/pdf/G20_policy_brief_Final.pdf. Retrieved: 22.03.2012.

Sutter, Christoph and Juan Carlos Parreño (2007). Does the current Clean Development Mechanism (CDM) deliver its sustainable development claim? An analysis of officially registered CDM projects. In: Climatic Change 84: 75–90.

Tabb, William (2011). The Restructuring of Capitalism in Our Time. Columbia University Press.

Taylor, Robert , Chandrasekar Govindarajalu, et al. (2008). Financing Energy Efficienty. Lessons from Brazil, China, India, and Beyond. Washington D.C.: World Bank.

The Brookings Institution (2011). China´s Low-Carbon Development. Confernce Proceedings. Washington, D.C. Tuesday, May 31, 2011.

The Climate Group (2009). Cutting the cost. The economic benefits of collobarative climate action. Cambridge. http://www.theclimategroup.org/_assets/files/Cutting_the_Cost_-_BTCD_Report.pdf. Retrieved: 22.03.201

Third World Network (2008). World Bank Climate Funds under fire from G77 and China. TWN Info Service on Finance and Development (Apr08/01), 7 April 2008. http://www.twnside.org.sg/title2/finance/twninfofinance200804 01.htm. Retrieved: 20.02.2010.

Thompson, Kevin (2010). Response to Colin Koopman´s "Historical Critique or Transcendental Critique in Foucault: Two Kantian Lineages. In: Foucault Studies 8: 122-128.

Tol, RSJ (2005). The marginal damage costs of carbon dioxide emissions: an assessment of the uncertainties. In: Energy Policy 33(16): 2064–2074.

Traue, Boris (2010). Das Optionalisierungsdispositiv: Diskurse und Techniken der Beratung, in: Johannes Angermüller and Silke van Dyk: Diskursanalyse meets Gouvernementalitätsforschung. Zur Einführung. Frankfurt am Main: Campus, 237-259.

Tsing, Anna (2004). Inside the Economy of Appearances, in: Ash Amin and Nigel Thrift: The Blackwell Cultural Economy Reader. Oxford: Blackwell, 83-100.

UK Committee on Climate Change (2008). Building a low-carbon economy – The UK's contribution to tackling climate change. The First Report of the Committee on Climate Change. London: The Stationery Office.

UK Department of Trade and Industry (2003). Our energy future - creating a low carbon economy. London: The Stationary Office.

UK Government (2008). Building a low carbon economy: unlocking innovation and skills. London: Department for Environment, Food and Rural Affairs.

UK Government (2009). The UK Low Carbon Industrial Strategy. London: Department for Business, Innovation and Skills/ Department of Energy and Climate Change.

Umweltbundesamt (2005). Klimawandel in Deutschland. Vulnerabilität und Anpassungsstrategien klimasensitiver Systeme. Dessau: Umweltbundesamt.

UN Economic Commission for Africa (2009). Carbon Finance in Africa - Extract. Africa Partnership Program. Addis Ababa, Ethiopia, 3 September 2009. http://www.oecd.org/dataoecd/45/46/43574330.pdf. Retrieved: 22.03.2012.

UNEP (1995). UNEP Statement of Environmental Commitment for the Insurance Industry. Geneva: United Nations Environment Programme.

http://www.unepfi.org/fileadmin/statements/ii/ii_statement_en.pdf.
Retrieved: 20.03.2011.

UNEP (1997). UNEP Statement by Financial Institutions on the Environment & Sustainable Development. Revised Version. Geneva: United Nations Environment Programme. Geneva: United Nations Environmental Program. http://www.unepfi.org/fileadmin/statements/fi/fi_statement_en.pdf. Retrieved: 20.02.2010.

UNEP (2009). Global Green New Deal: A Policy Brief: United Nations Environmental Program.

UNEP (2009a). Global Green New Deal. Policy Brief. Geneva: United Nations Environmental Programme.

UNEP (2010). Overview of the Republic of Korea´s National Strategy for Green Growth. Geneva: United Nations Environmenal Programme.

UNEP Finance Initiative (2009). Financing a global deal on climate change. Geneva: United Nations Environment Programme.

UNEP SEF Alliance (2009). Why Clean Energy Public Investment Makes Economic Sense - The Evidence Base. Basel: United Nations Environmental Programme.

UNFCCC (2005). Decision -/CP.11: Further Guidance for the operation of the Least Developed Countries Fund. unfccc.int/files/meetings/cop_11/application/pdf cop11_02_4e_ii_further_guidance_ldc_fund.pdf. Retrieved: 18.02.2011.

UNFCCC (2007). Investment and financial flows to address climate change. Bonn: UNFCCC.

UNFCCC (2007). Report on the analysis of existing and potential investment and financial flows relevant to the development of an effective and appropriate international response to climate change. Bonn: UNFCCC.

UNFCCC (2008). Report of the Conference of the Parties on its thirteenth session, held in Bali from 3 to 15 December 2007. Addendum, Part Two: Action taken by the Conference of the Parties at its thirteenth session. FCCC/CP/2007/6/Add.1.

UNFCCC (2008). Report of the Conference of the Parties on its thirteenth session, held in Bali from 3 to 15 December 2007. FCCC/CP/2007/6/Add.1. http://unfccc.int/resource/docs/2007/cop13/eng/06a01.pdf. Retrieved: 22.03.2012.

UNFCCC (2009). Copenhagen Accord. Proposal by the President. FCCC/CP/2009/L.7. Draft decision -/CP.15. UNFCCC.

United Nations (1992). United Nations Framework Convention on Climate Change. New York.

United Nations (1998). Kyoto Protocol to the United Nations Framework Convention on Climate Change.

Urban, Frauke and Wang Yu (2009). Climate Change, Energy, and Low-carbon Development in the Chinese Context. IDS Policy Briefing 8.2. Brighton: Institute of Development Studies.

Wara, Michael (2007). Is the global carbon market working? In: Nature 445(8): 595-59

Ward, John , Sam Fankhauser, et al. (2009). Catalysing low-carbon growth in developing economies. Public Finance Mechanisms to scale up private sector investment in climate solutions. Geneva: UNEP and Partners.

WBCSD (2009). Towards a Low-carbon Economy. A business contribution to the international energy & cllimate debate. Geneva: World Business Council for Sustainable Development.

WBCSD (2011). Scaling up low-carbon investment through the UNFCCC. Submission to the UNFCCC, 21 February 2011. Geneva: World Business Council for Sustainable Development.

WBGU (2007). Climate Change as security threat. Summary for Policy Makers. Berlin: Wissenschaftlicher Beirat Globale Umweltveränderungen.

WEF (2009). Task Force on Low-Carbon Prosperity. Summary of Recommendations. Geneva: World Economic Forum.

WEF (2011). Scaling-up Low Carbon Infrastructure Investment in developing countries. Geneva: World Economic Forum.

WEF (2011). Scaling-up Low Carbon Infrastructure Investment in Developing Countries. Geneva: World Economic Forum. Retrieved:

Weitzmann, M (2007). On modeling and interpreting the economics of catastrophic climate change http://www.economics.harvard.edu/faculty/weitzman/files/REStatModeling.pdf. Retrieved: 22.03.2012.

Weitzmann, Martin (2007). A Review of The Stern Review on the Economics of Climate Change. In: Journal of Economic Literature 45: 703-724.

World Bank (2006). Managing Climate Risk. Integrating Adaptation into World Bank Group Operations. . Washington D.C.: World Bank.

World Bank (2008a). Proposal for a strategic climate fund. Design Meeting on Climate Investment Funds. CIF/DM.2/3. Washington D.C.: World Bank.

World Bank (2008b). Strategic Climate Fund. Washington D.C.: World Bank.

World Bank (2008c). State and Trend of the Carbon Market 2008. Washington D.C.: World Bank.

World Bank (2008d). Carbon finance for sustainable development. Washington D.C.: World Bank.

World Bank (2009a). 10 Years of experience in Carbon Finance. Insights from working with carbon markets for development & global greenhouse gas mitigation. Washington D.C.: World Bank.

World Bank (2009b). Clean Technology Fund Investment Plan for Mexico. Document Prepared for the CTF Trust Fund Committee Meeting January 29-30, 2009, CTF/TFC.2/8. Washington D.C.: World Bank.

World Bank (2009c). Development and Climate Change. World Development Report 2010. Washington D.C.

World Bank (2009d). Clean Technology Fund. Guidelines for Investment Plans. Washington D.C.

World Bank (2009e). Clean Technology Fund. Investment Criteria for Public Sector Operations. Washington D.C.: World Bank.

World Bank (2009f). Clean Technology Fund. Financing Products, Terms, and Review Procedures for Public Sector Operations. Washington D.C.: World Bank.

World Bank (2009g). First "World Bank Green Bonds" Launched. Press Release, January 5, 2009. Washington D.C.: World Bank.

World Bank (2010a). CTF Financing Products, Terms and Review Procedures for Private Sector Operations. Washington D.C.: World Bank.

World Bank (2011). Green Bond Fact Sheet. Updated July 26, 2011. Washington D.C.: World Bank.

Wullweber, Joscha and Christoph Scherrer (2010). Post-Modern and Post-structural International Political Economy, in: Robert Denemark: The International Studies Encyclopedia. Oxford: Blackwell.

Yohe, Gary and Richard Tol (2008). The Stern Review and the economics of climate change: an editorial essay. In: Climatic Change 89: 231-240.

Young, Tom (2010). World Bank's green bonds clear $1bn mark. Bond issues raise finance for low-carbon and climate adaptation projects in developing world. http://www.businessgreen.com/bg/news/1801008/world-banks-green-bonds-usd1bn-mark. Retrieved: 5 September 2011.

Zadek, Simon (2011). Beyond climate finance: from accountability to productivity in addressing the climate challenge. In: Climate Policy 11(3): 1058-1068.

VS Forschung | VS Research
Neu im Programm Politik